A Guide to Measured Term Contracts

A Guide to Measured Term Contracts
Value for Money in Property Maintenance and Improvements

London: The Stationery Office

Applications for reproduction should be made in writing to Carillion Services Ltd, West Link House, 981 Great West Road, Brentford, Middlesex TW8 9DN.

The information contained in this publication is believed to be correct at the time of manufacture. Whilst care has been taken to ensure that the information is accurate, the publisher can accept no responsibility for any errors or omissions or for changes to the details given.

A CIP catalogue record for this book is available from the British Library
A Library of Congress CIP catalogue record has been applied for

Crown copyright material is reproduced with the permission of the Controller of HMSO and the Queen's Printer for Scotland

First published 2001
ISBN 0 11 702554 2

the
Stationery Office

Published by The Stationery Office and available from:

The Stationery Office
(mail, telephone and fax orders only)
PO Box 29, Norwich, NR3 1GN
Telephone orders/General enquiries 0870 600 5522
Fax orders 0870 600 5533

www.thestationeryoffice.com

The Stationery Office Bookshops
123 Kingsway, London WC2B 6PQ
020 7242 6393 Fax 020 7242 6394
68–69 Bull Street, Birmingham B4 6AD
0121 236 9696 Fax 0121 236 9699
33 Wine Street, Bristol BS1 2BQ
0117 926 4306 Fax 0117 929 4515
9–21 Princess Street, Manchester M60 8AS
0161 834 7201 Fax 0161 833 0634
16 Arthur Street, Belfast BT1 4GD
028 9023 8451 Fax 028 9023 5401
The Stationery Office Oriel Bookshop
18–19 High Street, Cardiff CF1 2BZ
029 2039 5548 Fax 029 2038 4347
71 Lothian Road, Edinburgh EH3 9AZ
0870 606 5566 Fax 0870 606 5588

The Stationery Office's Accredited Agents
(see Yellow Pages)

and through good booksellers

Printed in the United Kingdom by The Stationery Office

TJ005519 C 10/01 655716 19585

Contents

List of tables   *vii*

About Carillion Services Ltd   *ix*

1: Introduction   **1**

2: History   **2**

3: What is a Measured Term Contract?   **3**

Types of Measured Term Contract   4

Constituent elements of a Measured Term Contract   4

Alternatives to Measured Term Contracts   5

Getting advice   6

4: Pre-contract   **8**

Pre-tender stage   8

Documentation   11

5: The tendering process   **12**

The tender evaluation   12

6: Post-contract   **16**

The Initial Contract Meeting   16

Post-contract control   17

The Housing, Grants, Construction and Regeneration Act 1996   21

7: Advantages and disadvantages   **22**

Advantages   22

Disadvantages   23

8: Schedules of Rates   **24**

Introduction   24

Role of the Schedule of Rates   24

Contents of a Schedule of Rates   25

Attributes of a good Schedule of Rates   27

Types of Schedules of Rates   28

Availability of Schedules of Rates   30

Published Schedules of Rates available   31

Fluctuations   32

9: Standard Forms of Contract   **34**

Introduction   34

The information required from the employer   34

Details to be inserted by the Contractor 38

JCT Standard Form of Measured Term Contract 39

GC/Works/7 (1999) Measured Term Contract 48

10: Computerised Measured Term Contract administration 57

Introduction 57

Works Orders and Estimates 57

The Order Ledger 58

Estimating 58

Composite items 58

Why use computerised Schedules of Rates? 59

Choice of system 59

Specifying computer systems 59

Choice of operator 60

Advantages and disadvantages of computerisation 60

Conclusion 61

11: Audit of Measured Term Contracts 62

Introduction 62

Background 62

Risks associated with Measured Term Contracts 63

Audit procedures 65

Reviews and audits 68

Conclusion 71

12: The Contractor's view 72

Introduction 72

Estimating and tendering 72

13: Summary 79

The future 79

A new role for the Measured Term Contract 80

Appendix 1: GC/Works/7 Model Forms 82

Appendix 2: Useful contact addresses 101

Bibliography 103

Tables

Table 1.1   Construction output at constant 1995 prices   1

Table 2.1   Contracting methods for maintenance and minor new works in PSA, 1982–89   2

Table 3.1   Types of Measured Term Contracts used by the PSA   4

Table 5.1   Measured Term Contract, tenders received   13

Table 5.2   Measured Term Contract, tender evaluation – 1   14

Table 5.3   Measured Term Contract, tender evaluation – 2   15

Table 6.1   Example: Measurement and Valuation of a Works Order   20

Table 8.1   Choice of Schedule of Rates   29

Table 9.1   Comparison of Contract Forms   56

Table 12.1   Comparison of rates   74

Table 12.2   Possible weighting of work categories   74

Table 12.3   Possible totals   75

Table 12.4   Possible preliminaries, build-up   76

Table 12.5   Percentage calculation   76

Table 12.6   Addition for fixed price   78

About Carillion Services Ltd

Carillion Services has many years of experience in the production of technical publications, for use by construction professionals in both the private and public sectors. This derives from its origins within the Property Services Agency, where a comprehensive portfolio of publications was developed to meet the needs of the Government Estate.

This book was written as a result of many requests from clients and managers for guidance on the tendering for, and administration of, Measured Term Contracts as one of the methods of obtaining better value for money in maintenance and repairs to properties.

Carillion Services is a modern, forward-looking and quality-driven consultancy providing a wide range of services which include:

- Measured Term Contract advice and training
- bespoke schedules of rates and specifications
- financial, technical, procedural and value-for-money audits
- property management
- facilities management
- cost consultancy and quantity surveying
- building surveying
- building control and fire consultancy
- specialist contractual and claims advice
- arbitration and adjudication.

To find out more about Carillion Services' extensive back-up services, please contact:

Carillion Services Ltd
Westlink House
981 Great West Road
Brentford
Middlesex TW8 9DN

Tel: 0870 128 5220
Fax: 0870 128 5204

scheduleofrates@carillionplc.com

www.carillionplc.com

Whilst all reasonable care has been taken in the preparation of this publication, Carillion Services will not be under any legal liability of any kind in respect of any misstatement, error or omission contained herein.

Acknowledgement

Carillion Services have summarised the JCT Standard Form of Measured Term Contract 1998 (published by RIBA Publications, London, 2000) displayed in Chapter 9 of this book. Similarly, the GC/Works/7 (1999) Measured Term Contract (published by The Stationery Office, London, 2000) has been summarised and thence displayed in Chapter 9. The Model Forms found in Appendix 1 are taken directly from GC/Works/7 (1999) Measured Term Contract.

1 Introduction

The management of maintenance contracts is often regarded as a matter of little importance, mainly because the value of individual orders is very small. However, the annual, overall volume of maintenance work let by many organisations often outweighs their expenditure on new work. A more vigorous approach to the management and cost control of maintenance work is therefore necessary.

Table 1.1 Construction output at constant 1995 prices

£ million 1995 prices

| Year | Housing | Repair and Maintenance (R&M) | | | Total (all work) | R & M as a % of Total |
| | | Other work | | Total | | |
		Public	Private			
1989	17,148	6,541	6,223	29,912	58,137	51.45
1990	16,821	6,684	6,335	29,840	58,375	51.12
1991	14,809	6,033	5,809	26,651	54,133	49.23
1992	13,821	5,587	5,481	24,889	51,927	47.93
1993	13,661	5,232	5,390	24,283	50,980	47.63
1994	14,356	5,430	5,607	25,393	52,692	48.19
1995	14,595	5,398	5,978	25,971	52,643	49.33
1996	14,655	5,119	6,859	26,633	53,863	49.45
1997	14,970	4,826	7,221	27,017	55,468	48.71
1998	14,740	4,749	7,436	26,925	56,370	47.76
1999	14,263	4,690	7,330	26,283	56,903	46.19

Source: Table 10, 'Economic Significance of Maintenance: Maintenance Expenditure 1989–1999', Building Maintenance Information (BMI) Special Report 295

Measured Term Contracting is one of the methods of obtaining value for money in maintenance. It has distinct advantages over 'cost plus'-type contracts in that the price paid to the Contractor is defined in a Schedule of Rates. The Contractor carries the responsibilities of his own inefficiencies, there is less need for site supervision and, if administered properly, less opportunity for disagreement of final accounts.

The use of Measured Term Contracts (MTCs) has increased rapidly in recent years as government initiatives have led local authorities and others to introduce competitive tendering for maintenance work.

We will look at the nature of Measured Term Contracts, where they can be used, the documentation required, how they are operated and their advantages and disadvantages.

2 History

The Property Services Agency (PSA) and its predecessors managed the construction and maintenance of public buildings for over 600 years, but it was not until 1865 that the first known Schedule of Rates was published. The old War Department Schedule, which included ten trades and over 7,000 prices, was used in many corners of the world for maintenance of the Defence estate. Building and maintenance work could be valued at pre-determined prices leading to the gradual emergence of the Term Contract.

However, the most sweeping move towards Measured Term Contracts in the public sector came 100 years later, in 1965, when the newly formed Ministry of Public Building and Works took over responsibility for maintenance of the whole of the Defence works in addition to the Civil Estate. The Ministry recognised the advantage to be gained from a series of 'call-off' contracts or standing arrangements whereby a heavy programme of maintenance and small works could be ordered at pre-set prices without the need for raising a separate contract on each occasion.

Since then, the Measured Term Contract has become the cornerstone of the maintenance operation and, although it is not without problems, it will serve the user well if used appropriately.

The PSA's dependence on the Measured Term Contract is illustrated by the following statistics showing the average annual expenditure in £millions over the period 1982–89 for different types of contract.

Table 2.1   Contracting methods for maintenance and minor new works in PSA, 1982–89

Contract type	Annual spend £ million	%
Measured Term Contracts	285	40
Lump Sum Contracts (£25,000–£150,000)	160	23
Directly employed labour	95	13
Jobbing Contracts (not exceeding £5,000)	60	9
Lump Sum Contracts (under £25,000)	45	7
Stores	30	4
Specialist Term Contracts	25	3
Daywork Term Contracts	10	1
	710	100

Source: Property Services Agency

The enormous use that PSA made of the Measured Term Contract – amounting to £285 million per annum or 40% of the total spend – represents about 1,000 Measured Term Contracts countrywide. However, this method of contracting is just as useful whether the Employer has one contract or 1,000.

3 What is a Measured Term Contract?

The Measured Term Contract is an arrangement whereby a Contractor undertakes to carry out a series of Works Orders, over a period of years, within a defined geographical area and where the work is subsequently measured and valued at rates contained in a pre-priced Schedule of Rates.

In order to obtain competition, tenderers submit a Contract Percentage on or off the Schedule of Rates to reflect their assessment of all additional factors (such as overheads, market conditions, etc.). Separate percentages are sometimes requested for different value bands or categories of work.

The Contract Percentage is subsequently applied to all work priced against the Schedule of Rates. Where one of the PSA Schedules of Rates is used, a second adjustment is made – to update the pricing level of the Schedule of Rates to the current month, using published percentages issued by the Department of Trade and Industry (DTI).

A Measured Term Contract is appropriate in the following circumstances:

1. where the Employer has an ongoing need for maintenance/minor new work – the contract period is generally a one- to three-year term, although longer periods of up to five years have been known for specialist contracts, e.g., Landscape Management

2. where there is sufficient workload to offer continuity and economy – the workload should be large enough to attract Contractors, and to offer them continuity of work, which in turn should provide the Employer with a saving in cost and documentation

3. where a prompt response is required – the Measured Term Contract provides the opportunity for a quick response, but needs good communication and management to achieve a consistent response over a period of three years or more.

The user also needs to consider the availability of suitable Contractors and any particular requirements of the Employer.

The Measured Term Contract is essentially a package. It will not necessarily provide the lowest price for each individual Works Order, but it will usually produce the best overall solution to an Employers' maintenance needs.

The price paid is a classic case of 'swings and roundabouts'. If a Contractor has to replace all the rainwater goods on a housing estate, the Schedule of Rates will reimburse him handsomely; on the other hand, if he is required to travel 50 miles to replace a pane of glass, he will not be so fortunate. It is important, therefore, to order all work covered by the Schedule of Rates from the Contractor – not just the low-value and awkward work. The Contractor will have tendered on the basis of all relevant work with a spread of value.

Types of Measured Term Contract

The various Measured Term Contracts used nationally relate to the various Schedules of Rates available on the market. In the PSA, the Building and Civil Engineering Schedule of Rates was the largest, containing about 20,000 items in tabular form and accounting for over 40% of the Measured Term Contract workload. There are other Schedules of Rates widely used to cover the vast majority of maintenance and minor new work the user is likely to experience.

Table 3.1   Types of Measured Term Contracts used by the PSA

MTC type	Annual spend £ million	%
Building works	121	43
Decoration work	23	8
Grounds maintenance	33	11
Roadworks and paving	24	8
Electrical services	53	19
Mechanical services	28	10
Maritime services	2	1
Railway works	1	–
	285	100

Source: Property Services Agency

Constituent elements of a Measured Term Contract

The main constituent elements of a Measured Term Contract are:

- the Schedule of Rates
- the Contract Conditions
- the Particular Clauses.

The Schedule of Rates

A Schedule of Rates consists of a list of unit items of work priced at a rate per unit (e.g., per m^2, kg, etc.). The items of work are fully described and generally include specification clauses. Usually, Schedules of Rates do not include quantities. Dependant on the type of Schedule of Rates used, the list of items will either be pre-priced by the Schedule writer or priced by Contractors during the tendering process. The Schedule of Rates thus formed is used in conjunction with the measurement of work to calculate payment. Schedules of Rates are explored in greater detail in Chapter 8.

The Contract Conditions

The Contract Conditions are the 'small print' and define the rights and responsibilities of the two parties to the contract and such details as the responsibility for measurement and periods of payment. Some employers write their own contract conditions, but it is usual to use a Standard Form produced by institutions or organisations with a vested interest in contracts that will produce the minimum of disputes. Standard Forms have the advantage of familiarity, as many Contractors, consultants and Employers will have used them before. Users may also have developed established practices and procedures in line with Standard Forms. In addition, it is also obviously expensive to have a contract written from first principles using legally qualified staff. Standard Forms are explored in greater detail in Chapter 9.

The Particular Clauses

Particular Clauses usually relate to scope of the work, overtime, sub-contracting, and any Employer's, or other's, requirements and particular site circumstances such as:

- security arrangements
- service installations
- 24-hour/365 days per year emergency call-out service
- 24-hour/365 days per year on-site attendance service
- specialist sub-contractors
- minimum and maximum response times.

Alternatives to Measured Term Contracts

Measured Term Contracts are only one of a number of strategies suitable for procuring maintenance, repair and minor new work. Other strategies include Daywork Term Contracts, Lump Sum Contracts and Jobbing Contracts. Particular circumstances will dictate which is the most suitable. However, it is important to note that they can all be used in conjunction with each other.

Daywork Term Contracts

Daywork or 'Cost Plus' contracts can be used where there is a very large number of very small Works Orders and the overhead costs far exceed the cost of actually carrying out the work. For example, changing a fuse may take little more than a minute, but getting to the house or flat, gaining access and returning to the Contractor's base may take anything from a few minutes to an hour. The percentage adjustment to a Schedule of Rates required by a Contractor may be disproportionate to the rates. It may be that the only solution in such a case would be to reimburse the Contractor his costs plus tendered percentage adjustments. This would be in the form of labour at a rate per hour and the cost of materials and plant.

The use of Daywork or 'Cost Plus'-type contracts, however, brings high risks to the Employer and should only be used as a last resort. There is little incentive for the Contractor to be efficient, as he will be reimbursed no matter how long he takes to complete a Works Order. Use of a Measured Term Contract means that the Contractor carries the responsibilities of his own inefficiencies and, consequently, there is less need for supervision.

Lump Sum Contracts

The Lump Sum Contract is usually based on specification, and sometimes drawings, and has been the main alternative to the Measured Term Contract in the public sector for many years. It was intended for use where specialist work, or work that could not be covered by the Schedule of Rates, was involved but it has been used by some as the main method of procuring maintenance work. The invitation of tenders for every job obviously provides competitive prices, but the paperwork and administration costs can be enormous.

Jobbing Contracts

A Jobbing Contract is where a Form of Tender is sent to the Contractor with a description of the work required, or with a specification, and/or drawings of the work. The Contractor completes his tender, and signs and returns it. The Employer countersigns the tender to accept it, and sends a copy back to the Contractor to complete the contract.

It is also possible to order work over the telephone, without the price or contract period being set, followed by a written order (provided that the time and price are reasonable). It is assumed that the Employer has an agreed list of contractors who have previously been provided with the Conditions of Contract and that issues such as insurance, etc. have been agreed.

Standard Conditions of Contracts are available, such as the 'Standard Form of Tender and Agreement for Building Works of a Jobbing Character' (JA/T90, JCT, 1990) produced by the Joint Contracts Tribunal. These can be used with a Standard Form of Tender, or with an Employer's own Works Order. Many Employers, particularly those in the public sector, issue their own Works Orders and, provided that they contain the necessary information, they can be used with the Conditions of Contract.

The information required is as follows:

- the start and finish dates
- the minimum amount of public liability insurance
- defects liability period – this is stated as six weeks, but can be changed if required
- the name of the Employer's representative.

Getting advice

When a problem of bad value for money on a site with a high annual spend is identified, it is necessary to analyse the methods of procuring maintenance and minor new works, the contract strategy, and the probity of procedures. The procurement and control of maintenance works are issues that building owners need to consider on a regular basis. To be certain that maintenance is properly managed, building owners need to establish whether works are being planned, procured and controlled in the most appropriate way.

Property Management Performance Evaluation

During the life of a building, the maintenance and running costs can amount to more than twice the initial capital cost of construction. However, small improvements in the management of maintenance can, amongst other things, produce significant financial savings over the life of the property portfolio. Property Management Performance Evaluations should be conducted by appropriately qualified professionals experienced in their particular discipline (generally building surveyors and quantity surveyors), and the teams should be structured to the particular needs of the evaluation. They should also be – and perceived to be – totally independent. There must be assurance that the evaluation is conducted objectively. There can be no accusations of bias or vested interests.

Measured Term Contract Auditing

Even when Measured Term Contracts are in place, it is advisable to implement a programme of regular audits. Employers will be seeking a measure of assurance that Measured Term Contract accounts accurately reflect the value of work properly ordered and executed; in other words, that they are in accordance with the contract. This should be developed to take the following factors into account: assurance, deterrence, value for money, quality of work executed and cost effectiveness.

- *Assurance* – There is a need to confirm that Measured Term Contract accounts are within tolerable levels of accuracy. The aim is to advise and assure the Employer on the level of accuracy of Measured Term Contract accounts so that action can be taken if there is any apparent cause for concern.

- *Deterrence* – The aim is to seek to deter those who are responsible for measuring on Measured Term Contracts from misrepresenting the amount properly due on any account, either through inefficiency or with intent to defraud. In order to establish deterrence, it should be widely known that checking is carried out. Operatives will not be deterred if there is no risk of being found out. In addition, the Works Orders to be checked must be unknown to those who measure them. Firm action will need to be taken against those found to be in serious default, and potential defaulters must be made aware of the consequences of their actions.

- *Value for money* – The Employer must be confident that Measured Term Contracts are being appropriately used. Other methods of procurement, usually lump sum tendering, may sometimes be more appropriate and should not be ignored. On the other hand, the Contractor will rightly expect a fair balance of workload for the Measured Term Contract to remain viable to him.

- *Quality of work executed* – A Measured Term Contract account should reflect only the quality of work actually provided by the Contractor, which could sometimes be inferior to that ordered. Auditors should take account of quality of work in the re-measured value of an audited account.

- *Cost effectiveness* – Experience has shown that a properly structured audit procedure will result in a reduction of overpayments on Measured Term Contract accounts. The statistics derived from analysis of the various checks undertaken show that the amount spent on audit is less than the total overpayment that would otherwise be incurred.

The auditing of Measured Term Contracts is explored in greater detail in Chapter 11.

4 Pre-contract

Pre-tender stage

Contract documents will normally consist of a Schedule of Rates, a Form of Contract and Particular Clauses giving project-specific details. An assessment of the anticipated workload will need to be made, together with the availability of finance, current market conditions and suitability of local Contractors. If there is an existing Measured Term Contract in place, the temptation to use the same documentation should be resisted. Standard Forms of Contract and published Schedules of Rates are constantly being revised and new editions produced. Local site conditions may have changed in respect of, e.g., scope of work, response times, access to site and security.

Here, following, is a summary of the information and documentation that is required, or needs consideration, at the pre-tender stage.

Schedule of Rates

First and foremost the user needs to decide which Schedule of Rates to use. The Building Employers' Confederation has recommended that: 'Local authorities should be required to use a uniform Schedule of Rates (such as the PSA Schedule of Rates or the National Schedule) that private Contractors will recognise and trust.' Schedules of Rates are discussed in greater detail in Chapter 8.

Fluctuations

Measured Term Contracts can have a contract period of one, two, three or even five years. One-year contracts are uncommon as the set-up costs for the Contractor can be high; only after three to five years is he gaining a significant advantage from continuity. However, if the contract period is of this duration, there should be a fluctuation price condition in the contract. Fluctuations can be allowed for by the application of an index on a monthly basis as with the Updating Percentage Adjustments for Measured Term Contracts used with the PSA Schedules of Rates. Alternatively, they can be allowed for by annual re-pricing of the base schedule as with the National Schedule of Rates.

Location drawings

These are only necessary to define the extent of the estate or the geographical area to be covered by the Measured Term Contract. Frequently, the Contract area itself may be divided into separate locations, which also allows the tenderer to quote different percentages.

Contract Conditions

The Contract Conditions define the rights and responsibilities of the two parties to the contract. Some employers write their own Contract Conditions, but it is usual to use a Standard Form produced by institutions or organisations with a vested interest in contracts that will produce the minimum of disputes. Standard Forms have the advantage of familiarity as many Contractors, consultants and Employers will have used them before. Standard Forms are explored in greater detail in Chapter 9.

There are two major Standard Forms of Contract suitable for Measured Term Contracts:

1. the JCT Standard Form of Measured Term Contract 1998 (London, RIBA Publications, 2000)
2. GC/Works/7 (1999) Measured Term Contract (Property Advisers to the Civil Estate Central Advice Unit, London, The Stationery Office, 2000).

Particular Clauses

These 'preliminary'-type clauses include 'scope of the work', overtime, sub-contracting and any Employer's, or other's, requirement relating to the particular site (such as security arrangements or service installations).

Estimated Annual Value (EAV)

The assessment of the anticipated workload and the availability of finance can be used to estimate the total annual expenditure on the Measured Term Contract. This is generally known as the Estimated Annual Value, or 'EAV'. The Employer needs to know this for budgeting purposes, while the Contractor needs to know for planning of his resources and for tendering. It is important, therefore, to insert a realistic value based on historical data and current needs and to monitor the value of Works Orders against the EAV during the year. Failure by the Employer to meet the Estimated Annual Value is often one of the reasons given for termination by the Contractor.

Maximum and minimum order values

The rates and prices required by the Contractor will depend on the number and value of Works Orders issued. Large numbers of very small Works Orders are much more expensive to carry out than small numbers of large Works Orders. For this reason, it is often advisable to set a minimum Works Order value to compensate the Contractor for Works Orders that are so small as to be not cost effective.

Maximum Works Order values are also advisable. Over a certain value, the Employer will obtain better value for money from competitive lump sum tenders. However, in fairness to the MTC Contractor, this must also be stated as it will also affect his tender. The actual figures will vary dependent upon the type of contract and likely scope of work, but minimum Works Order values can vary from £NIL to £5,000, and maximum Works Order values vary from £10,000 up to £50,000.

Value bands

The Measured Term Contract should give the Contractor the opportunity to reflect the differing costs for varying sizes of Works Orders in his tendered prices. The usual method is by the use of 'value bands'. The industry has had an on-off relationship with value bands but now tends towards use of three bands: up to £5,000, £5,000–£25,000, and over £25,000.

Some Contractors tender the same percentage for all three bands while others show a marginal difference.

The system is flexible and can be adjusted to suit a variety of circumstances. For example, separate percentages may be requested on Painting Measured Term Contracts for work in different types of building such as occupied housing, unoccupied housing, airfield hangers, or telephone exchanges.

Daywork

There is inevitably going to be work that is not capable of being measured and valued against the Schedule of Rates. The usual remedy for this situation is the inclusion of a Daywork provision. Daywork is an expensive alternative to using contract rates: with work measured and valued against rates, it is in the Contractor's interest to complete as quickly as possible, to keep his costs within the defined sum he will receive; using Daywork, this incentive is negated.

When published Schedules of Rates do not cover particular regularly occurring work, consideration should be given to producing a local addendum.

If a Daywork provision is included, tenderers should be asked to quote Daywork rates using a prepared Daywork Schedule (see Model Form 4, in Appendix 1), and an appropriate definition should be included in the documentation.

Contract period

The choice of contract period is flexible. For example, a three-year period with an option to increase it to five years could be used for grounds maintenance. This arrangement enables full account to be taken of the seasonal nature of the work, the capital investment involved for the Contractor and the cost benefit to the Employer of a longer tenure by the Contractor. The system is very adaptable and the duration of terms can vary to suit the specific circumstances relating to particular types of work. A two-year term, for example, might be considered more appropriate for building works. Where circumstances are favourable, extension of contract periods can also be negotiated.

Termination should be allowed by either party, without penalty. Three-months' notice after a minimum of six months of the Contract period is usual.

Lump Sum Maintenance Element (LSME)

Opinion is divided on this subject. Measured Term Contract tenderers are sometimes invited to quote an annual lump sum for all maintenance work to a specific area. For example 'Repairs to Family housing', 'Boiler servicing' or 'Grounds Maintenance of Sportsfields'.

As the work is not 'measured' it does not have to form part of a Measured Term Contract, but if used needs very careful consideration. The advantages and disadvantages are a subject in their own right.

Quantity Surveying appointment and self measure

In order to agree the value of work carried out, the responsibility for measurement after the work has been completed needs to be established.

In an effort to reduce the cost of joint measurement of low-value Works Orders, it has become increasingly common to require the Contractor to measure and submit accounts for Works Orders up to £2,500. The implication on checking needs to be considered if this approach is used. It has been found that for work of this value, the saving in measurement costs is almost offset by the additional checking required, the higher level of over-measurement involved and the subsequent need for further checking. In some instances, it is cheaper to measure jointly – in other cases, there is a saving.

If the decision is made not to measure jointly, a system of random checking should be established. This is commonly about 5% of Works Orders measured by the Contractor. The Works Orders should be selected randomly but there should be provision for a minimum number of Orders covering activities with a high risk of fraud, e.g. Grounds Maintenance, Daywork, etc.

In the case of joint measurement, the appointment of a Quantity Surveyor who is experienced in measurement of Measured Term Contracts should be considered and the Contractor notified accordingly.

Documentation

The documentation is simple – no Bills of Quantities or Specifications to produce – just a set of standard documentation and careful thought in their adaptation to specific needs. The GC/Works/7 (1999) Measured Term Contract contains a number of standard forms that users unfamiliar with Measured Term Contracts may find useful; Model Forms 1–10 are featured in Appendix 1 of this book.

The following is a checklist of documentation that may be required:

✔ *Abstract of Particulars (see Model Form 1) giving basic Project information such as:*
  - the Employer
  - the Project Manager (PM) or Contract Administrator (CA)
  - the Contract Period
  - the Contract Area
  - the Maintenance Period
  - the Minimum and Maximum Order Values.

✔ *Invitation to Tender (see Model Form 2)*

✔ *Tender Form (see Model Form 3)*

✔ *Contract Daywork Schedule (see Model Form 4)*

✔ *Adjudicator's Appointment (see Model Form 5)*

✔ *Works Order Form (see Model Form 6)*

✔ *Interim Payment Certificate (see Model Form 7)*

✔ *Final Payment Certificate (see Model Form 8)*

✔ *Notice of Intention to Withhold Payment (see Model Form 9)*

✔ *Employer's Notice of Determination (see Model Form 10).*

5 The tendering process

Once the decision to proceed with a Measured Term Contract has been made and the documentation has been considered, the next stage is to appoint a Contractor. It is becoming increasingly common for Employers to adopt 'partnering' principles and, accordingly, they may already have a long-term relationship with a Contractor who they may wish to appoint to carry out the work. The usual procedure, however, is for competitive tenders to be invited. Many organisations with a property portfolio large enough to merit a Measured Term Contract have their own contracts procedures, but the basic principles are as follows:

1. The various constituent parts of the Measured Term Contract documentation are brought together to produce a clear and comprehensive tender invitation package that clearly reflects the scope of work and contract conditions.

2. A short list of suitable tenderers is drawn up, either from the Employer's list of approved Contractors or from other reliable sources such as recommendations from other organisations, submission of questionnaires, interview, etc. The selected tenderers should have proven competence in Measured Term Contracts of the size and nature under consideration as well as general contracting and administrative skills, integrity and responsibility. The number of tenderers selected to be invited to tender is open to choice, but it should be noted that, the higher the number, the lower the chances of any one tenderer being successful and this may be reflected in tender prices. Five or six tenderers is the usual choice.

3. Tenders should be submitted in sealed envelopes, preferably using pre-addressed labels without the Contractors' identity being discernible from the outside. The latest time for submission should be specified as an hour of the day on a particular date.

4. The submitted tenders are evaluated to find the lowest acceptable tender and the tender accepted in writing.

The tender evaluation

In the simplest case, the tendered percentage is merely applied to the Estimated Annual Value (EAV) and the Daywork percentages are applied against estimated sums. In practice, there may well be various percentages quoted for different value bands, different locations or, as in the case of electrical work or grounds maintenance, for separate sections of the Schedule of Rates. Therefore, a breakdown of the EAV is essential and this should always be calculated prior to invitation of tenders and kept securely.

Table 5.1, following, shows a typical tender return. Selecting the lowest tenderer merely by inspection would be a risk. Contractor 2 has tendered lowest for the 'not exceeding £5,000' value band, but highest for the 'over £25,000' band. Contractor 1 is marginally lower than Contractor 3 in the 'not exceeding £5,000' value band, but higher for the 'over £25,000' band.

Table 5.2, following, shows a tender evaluation. An Estimated Annual Value of £500,000 has been divided up and allocated to the various value bands and to sub-contractors/suppliers and daywork (column A). Estimated percentage adjustments, perhaps based on previous contracts, have been applied to the Estimated Annual Values (column B) to produce net values (column C). For example, the estimated value of Works Orders in the 'not exceeding £5,000' value band is 75% giving a value of £300,000 and a net value (£300,000 divided by 1+25%) of £240,000. The tendered percentages of each tenderer can now be applied to these net values to produce nominal tender sums (columns D, E and F) making comparison much easier.

Table 5.3 shows the importance of allocating percentages to the various value bands with as much accuracy as possible. In Table 5.2, Contractor 2 was the lowest. In Table 5.3, with different allocations, Contractor 2 is now the highest and Contractor 3 is the lowest. The opportunity to manipulate the figures to produce the desired effect is an obvious risk here and this is why it is highly recommended that the allocations are made prior to the invitation of tenders and kept in a secure location.

Table 5.1 Measured Term Contract, tenders received

	Tenders		
	Contractor 1	Contractor 2	Contractor 3
1.0 Measured work			
Not exceeding £5,000	22.0%	12.0%	23.0%
£5,000 to £25,000	7.5%	12.0%	5.0%
Exceeding £25,000	−13.0%	12.0%	−15.0%
2.0 Dayworks			
Labour			
Craftsmen	£12.00	£13.75	£11.80
Labourers	£9.10	£11.00	£9.25
Materials	12.5%	15.0%	25.0%
Plant	10.0%	15.0%	18.0%
3.0 Nominated sub-contract work	15.0%	12.5%	15.0%

Table 5.2 Measured Term Contract, tender evaluation – 1

Estimated Annual Value £500,000

	A		B	C	D	E	F
						Tenders	
Breakdown	% of EAV	Gross value	Estimated % adjustment	Net value	Contractor 1	Contractor 2	Contractor 3
1.0 Measured work	**80**	**£400,000**					
Not exceeding £5,000	75	£300,000	25.0%	£240,000	22.0% £292,800	12.0% £268,800	23.0% £295,200
£5,000 to £25,000	15	£60,000	5.0%	£57,143	7.5% £61,429	12.0% £64,000	5.0% £60,000
Exceeding £25,000	10	£40,000	–10.0%	£44,444	–13.0% £38,667	12.0% £49,778	–15.0% £37,778
2.0 Dayworks	**10**	**£50,000**					
Labour	55	£27,500					
Craftsmen		65% £17,875	£14.50	1,233 hrs.	£12.00 £14,793	£13.75 £16,950	£11.80 £14,547
Labourers		35% £9,625	£10.50	917 hrs.	£9.10 £8,342	£11.00 £10,083	£9.25 £8,479
Materials	35	£17,500	15.0%	£15,217	12.5% £17,120	15.0% £17,500	25.0% £19,022
Plant	10	£5,000	7.5%	£4,651	10.0% £5,116	15.0% £5,349	18.0% £5,488
3.0 Nominated sub-contract work	**10**	**£50,000**	5.0%	£47,619	15.0% £54,762	12.5% £53,571	15.0% £54,762
Totals	**100**	**£500,000**			**£493,029**	**£486,031**	**£495,276**

Table 5.3   Measured Term Contract, tender evaluation – 2

Estimated Annual Value **£500,000**

		A	B	C	D		E		F	
					Tenders					
Breakdown	% of EAV	Gross value	Estimated % adjustment	Net value	Contractor 1		Contractor 2		Contractor 3	
1.0 Measured work	**80**	**£400,000**								
Not exceeding £5,000	35	£140,000	25.0%	£112,000	22.0%	£136,640	12.0%	£125,440	23.0%	£137,760
£5,000 to £25,000	40	£160,000	5.0%	£152,381	7.5%	£163,810	12.0%	£170,667	5.0%	£160,000
Exceeding £25,000	25	£100,000	–10.0%	£111,111	–13.0%	£96,667	12.0%	£124,444	–15.0%	£94,444
2.0 Dayworks	**10**	**£50,000**								
Labour	55	£27,500								
Craftsmen		65%		1,233 hrs.	£12.00	£14,793	£13.75	£16,950	£11.80	£14,547
Labourers		35%		917 hrs.	£9.10	£8,342	£11.00	£10,083	£9.25	£8,479
Materials	35	£17,500	15.0%	£15,217	12.5%	£17,120	15.0%	£17,500	25.0%	£19,022
Plant	10	£5,000	7.5%	£4,651	10.0%	£5,116	15.0%	£5,349	18.0%	£5,488
3.0 Nominated sub-contract work	**10**	**£50,000**	5.0%	£47,619	15.0%	£54,762	12.5%	£53,571	15.0%	£54,762
Totals	**100**	**£500,000**				**£497,250**		**£524,004**		**£494,502**

6 Post-contract

The Initial Contract Meeting

The Employer's team

The administration of a Measured Term Contract may require a number of staff to cover the various duties. Although the Project Manager (PM) or Contract Administrator (CA) retains responsibility under the contract, he will usually require a Quantity Surveyor (QS) to assist. The duties of the Quantity Surveyor should not be limited to post-contract measuring and valuing work ordered but should extend to advising on the cost implications of alternative materials or construction methods.

In practice, the Project Manager or Contract Administrator does not usually issue all Works Orders. There is often a site team responsible for issuing some Works Orders, keeping records and payment of bills.

The Contractor's team

The contract will usually require the Contractor to supply a competent agent to supervise the execution of the work, but a good Contractor will ensure that he has a team working on the contract dedicated to keeping the Employer happy. With the increasing prevalence of self-measurement, the Contractor will also require his own Quantity Surveyor.

The Initial Meeting

After formal acceptance of tenders, the contract should be started with an Initial Contract Meeting attended by the Employer's contract management team and the Contractor. The Agenda for the meeting could be as follows:

1. *Introductions*
   Record the names and telephone numbers of the attendees.

2. *The scope of the contract*
   Record the agreed start date and the Estimated Annual Value (EAV).

3. *The Employer's team*
   Record the names and telephone numbers of the Project Manager/Contract Administrator, Quantity Surveyor and Technical Staff.

4. *The Contractor's team*
   Record the names and telephone numbers of the Contractor's agent, Quantity Surveyor and Technical Staff.

5. *Anticipated workload*
   Record the current known workload including commencement, completion dates, urgency, response times, etc.

6. *Works Ordering*
   Approval of sub-contractors and materials, access to site, hours of work, site restrictions, responsibility for issuing Works Orders.

7. *Quantity Surveying*
   Responsibility for measuring, payments on account, final accounts.

8. *Health and Safety*
   Site hazards, access approvals, H&S legislation.

9. *Date of next meeting.*

It is important that a realistic contract start date is agreed. The Contractor will need a 'lead-in' period during which he can arrange his site and off-site organisation, technical staff, plant and material sources.

Post-contract control

Post-contract control will benefit from planning of the maintenance programme. It requires a disciplined approach with experienced and committed staff and good record-keeping. The anticipated workload and flow of work throughout the whole contract should be established by formulating a 'Forward Maintenance Plan' (FMP). This should take into account as many factors as possible including the current condition of the properties, proposed improvements, any new work and the availability of finance. Projections based on expenditure in previous years may prove useful.

The FMP can then be used to form an annual budget for the contract. As the final values of Works Orders are agreed, the information should be fed into the FWP to ensure the budget is constantly updated.

The following is a summary of other post-contract matters that need to be considered.

Ordering of Work

As soon as the Contract is let, a standard Works Order form giving a detailed specification, location and date for completion may be issued to instruct the Contractor to carry out work (for example, see Model Form 6 in Appendix 1). The workload should be carefully planned to maximise efficiency and value for money: small, similar Works Orders can be combined, for example. To improve response times, the Contractor should be permitted to have a small 'bank' of Works Orders to cover varying delivery times for materials. On the other hand, an excessive backlog will produce the opposite effect. A balance must be achieved.

Identification of requirements – It is important that work ordered under a Measured Term Contract is clearly identified and comes within the scope of work envisaged by the contract. Any special requirements of the contract, e.g., emergency call-out facility, and how it will be valued, should be clearly identified within the contract.

Design and specification responsibilities – The Schedules of Rates are compiled on the basis that work ordered is pre-designed and specified. Where a Contractor is required to undertake diagnostic work, to determine the work required, an additional charge to the measured value is to be expected.

Procurement method – Before ordering work, consideration should be given to whether the work envisaged is appropriate to the Measured Term Contract. For example, where the bulk of the work is not measurable against the Schedule of Rates, it may be better to procure the work a different way; e.g., lump sum quotations in competition.

Information required – All Works Orders should contain the following information:

- what – the specification and design of the work
- where – the location of the work (site, building, room, floor, elevation, etc.)
- when – the date by which the Contractor is expected to finish the work ordered
- how much – the extent of the work (e.g., first five near-lights at north end of corridor).

'Faulty light on first floor' is a wholly inadequate description. There may be many rooms on the first floor, which would require the Contractor to seek out the right one. Even when he has found the faulty light, he will need to establish what the fault is. Has the fluorescent tube failed, or has the switch been broken? The Contractor's time wasted tracking the fault is likely to be charged to the Employer in the form of Daywork. The work must be adequately described, such as 'Replace fluorescent tube Type A in room 6, first floor', and a date for completion stated.

The Schedules of Rates are compiled on the basis that work ordered is pre-designed and specified. Where a Contractor is required to undertake diagnostic work to determine the work required, an additional charge to the measured value is to be expected, e.g., 'Remedy radiator not working' would involve the Contractor determining 'why'. His time in identifying the problem is in addition to the measured value of the work.

Alternatively, 'Take out defective 20mm-diameter copper alloy radiator valve (angle pattern) and renew' specifies the work and is directly measurable against the Schedule of Rates.

Progress meetings

Regular progress meetings to monitor progress should be organised, usually monthly, and the Contractor viewed as one of the team. A 'firm but fair' approach to Contractors is required. Prior to the meeting, a current financial statement should be prepared covering the number of Works Orders issued, the value to date, Works Orders outstanding and Final Accounts outstanding. The Agenda for the meeting could be as follows:

1. *Attendance* (and apologies for absence)
   Record the names of the attendees.
2. *Approval of the Minutes of the previous meeting*
3. *Matters arising*
4. *Review of Works Orders*
   - Works Orders to be issued
   - Works in progress
   - Works completed but not valued
   - final accounts completed.
5. *Lump sum elements* (if any)
   Response, method of payment and quality of work.
6. *Record of labour on site*
7. *Other Works on site*
8. *Liaison with occupants*
9. *Date of next meeting.*

Management of information

The ordering system must be kept simple or the volume of paperwork can become overwhelming. A computer system should be used if possible. It is essential that, at any point in time, the total anticipated commitment, i.e., the total of Works Orders issued and the total of programmed work still to be ordered, is known. The consequences otherwise may be either an under-spend or over-spend, neither of which is desirable to the Employer.

Various computerised Measured Term Contract administration systems are available that record every Works Order issued, the type of work, Contractor, value and whether it has been checked or not. This will cover the production and valuation of all Works Orders, and significantly reduces the Quantity Surveying input. Computerised Measured Term Contract systems are explored in greater detail in Chapter 10.

Estimating

Why is an estimate necessary for each Works Order when the work will shortly be measured and valued at pre-set prices?

The main reason is budgetary control. The Employer needs to monitor budget and expenditure, and frequently the time-lag is too long between issue of a Works Order and agreement of its final account to do this effectively without an estimate.

Estimating is therefore necessary, but notoriously difficult, for small-value maintenance work. To assist technical staff, consideration should be given to the use of estimating aids including computer software in addition to the Schedule of Rates itself.

Advances on account

Under certain conditions stated in the Conditions of Contract, advances on account are payable, at monthly intervals, on any Works Order estimated to be over a designated value, and are usually based on a percentage assessment by the supervisory staff.

Supervision

Good supervision is vital. There has been a tendency to view maintenance work as the poor relation to new work, which is unwarranted as the overall spend is often as high and the need to supervise and to sign-off work as complete is just as relevant.

Although management controls are important, care must be taken not to weigh down supervisory staff with 'paperwork and procedures' at the expense of their main function of on-site supervision.

Measurement and valuation

Whether a Quantity Surveyor will measure and value all Works Orders, or whether the Contractor will self-measure Works Orders of a certain value will have been decided.

In either case, the description of work and dimensions are taken on site or from drawings. Schedule of Rates items are added and priced, usually in the office. The normal pricing rules apply. Rates used are:

- from, or based upon, the Schedule of Rates
- agreed rates, or
- Daywork.

A typical collection sheet will show the collected value of the measured work based upon the Schedule of Rates and then the two adjustments – firstly, for updating to the month of the Works Order, and secondly, adding or subtracting the contract percentage.

Daywork, or work at agreed rates, invoices and the like are then included but obviously do not attract the various percentages. Many quantity surveying firms design their own proformas for this summary sheet.

Table 6.1   Example: Measurement and Valuation of a Works Order

Measured Term Contract Order No: EXAMPLE 1

Measurement and Valuation

	Dim.	Tot.	Description	Item ref.	Unit	Qty.	Rate	Extension
			Frame					
	0.90	0.90	Dr fr/ hw/ 100 × 75mm					
			(nom)/ teak	L239/3	m		1.06	
2.00	2.10	4.20		L239/4	m		8.40	
		5.10					9.46	
				L1	factor		1.05	
			Finished size 94 × 69mm =					
			6486mm^2	K292	factor			2.50
						5.10	24.83	126.65
			L239/4					
			(6486 - 500)/100 =					
			60 × 0.14 = £8.40					
			Labours					
2.00	5.10	10.20	Rebate	L246/2	m		0.50	
				K292	factor		2.50	
						10.20	1.25	12.75
			Fixing					
	5.10	5.10	Plugging	P145	m	5.10	0.80	4.08
			Screwing and pelleting	P142/2	m		0.98	
				P144/2	m		2.00	
							2.98	
				K292	factor		2.50	
						5.10	7.45	38.00

Total				181.48
Add updating % May 2000			23%	41.74
Total (updated)				223.22
Deduct Contract %			−10%	-22.32
Net total (measured work)				**£200.90**

The Housing Grants, Construction and Regeneration Act 1996

Scope

The Housing Grants, Construction and Regeneration Act 1996 (amended 2001, London, The Stationery Office) applies to all construction contracts and includes not just construction operations themselves, but also the arrangement of the execution of construction operations by others, e.g., sub-contractors. The main provisions are in connection with:

- payment
- adjudication.

Payment – The Act stipulates a number of entitlements that must be incorporated into construction contracts. For example, if the duration of the work exceeds 45 days, the Contractor is entitled to interim payments. There must also be an adequate mechanism for determining when payments are due and how much.

Adjudication – The Act also introduces Adjudication. Arbitration is an established method of dispute resolution and it may seem unnecessary to introduce a new form. However, arbitration or legal proceedings can be very time-consuming and adjudication is designed to speed up the process.

7 Advantages and disadvantages

Advantages

Flexibility/Accountability

The cost of work can be evaluated before a Works Order is raised thereby providing a reliable and accurate estimate. Expenditure and commitment is easily monitored providing close financial control. There is no more flexible method of procuring small works, while retaining accountability, whether it is to the Public Accounts Committee, the taxpayer or the shareholders.

Minimal documentation

A multitude of Works Orders may be raised under the one contract with very little in the way of documentation. Tender action is only required once each term.

Immediate response

Without the requirement for contract action, the Contractor's response should be immediate once a Works Order is issued.

Saving in time and resource cost

Resulting from pre-contract documentation and post-contract response time.

Contractor 'on call'

Apart from being in a position to respond quickly to normal demands, the Contractor should be able to deal with any emergency work that is needed.

Familiarity/Continuity

The Employer benefits from the Contractor's increasing familiarity with the buildings, while the Contractor benefits from the assurance of a reasonable continuity of work. Once given a selection of Works Orders, the Contractor can programme his work to make the most efficient use of his resources.

There is an excellent opportunity for both parties to create a good, long-term working relationship.

Value-for-money

Measured Term Contracts provide extremely good value-for money. The PSA carried out an exercise of re-pricing over 100 Lump Sum Contracts (up to £25,000 in value) on the basis of available Measured Term Contracts. Results showed that overall, taking account of 'swings and roundabouts', there was little or no difference in pricing levels. There may be a potential saving by use of Lump Sum Contracts for the higher-value Works Orders but the lower-value work – which forms the bulk of Measured Term Contract spend – showed a clear advantage to the term contract.

Termination

The termination clause is fair to both parties offering a remedy for poor performance or a change in circumstances.

Simplicity

Tender documentation, adjustment for inflation and the ordering of work are all simple operations.

Disadvantages

Volume of measurement

Frequently there is a comparatively large amount of measurement for the small-value Works Order – hence the move towards self-measurement, which may itself require closer control.

Delayed accounts

There is sometimes insufficient incentive for the Contractor and Quantity Surveyor to be up-to-date with measurement and valuation, particularly where the Contractor has received advance payments. This situation has been recognised and the latest Contract Conditions address the problem. Nevertheless, regular monitoring is recommended.

Familiarity/convenience

Once a term contractor is established, it becomes very convenient to use him for work outside his remit and this can overtake commercially sound reasons to adopt other procurement methods.

Availability of suitable contractors

Generally, term contracts require Contractors with considerable all-round expertise and this can limit the availability of suitable Contractors.

Audit requirement

The volume of separate Works Orders and high value of Measured Term Contract spend does create an auditing problem.

8 Schedules of Rates

Introduction

The Schedule of Rates is the key component in a Measured Term Contract as it is used to determine the successful tenderer and forms the contractual basis on which payment is made. Schedules of Rates are also used in conjunction with Lump Sum Contracts to provide a basis for valuing variations to the contract, but this is a separate topic in its own right. They can also provide a benchmark for comparing regional variations and monitoring tendering trends

This chapter describes the role of the Schedule of Rates within a Measured Term Contract, the contents of a Schedule, the types of Schedule available and the choice and preparation of Schedules. It is worthwhile at this stage to make the distinction between 'Schedules of Rates' and 'Price Books'.

There are many Price Books available but these are used, primarily, for estimating building works. They could, theoretically, be used as Schedules of Rates, but published Schedules have the advantage of the inclusion of measurement rules and specification notes, the availability of compatible computerised administration systems and historical links with published Standard Forms of Contract.

It is important to the success of a Measured Term Contract that an appropriate Schedule of Rates is provided. Time spent in choosing or preparing a Schedule of Rates to suit particular circumstances will be more than repaid in savings during the operation of the Contract.

Role of the Schedule of Rates

The Schedule of Rates fulfils the same two primary roles as a Bill of Quantities fulfils for a Lump Sum Contract, i.e., it provides the basis for tender assessment and the basis for payment. It also fulfils some secondary roles in facilitating budget planning, cost control and ordering.

Primary functions

Defining specification standards – A Schedule of Rates can be prepared to include materials and workmanship specification clauses. These are either incorporated in item descriptions or separate preamble clauses. Although the Project Manager/Contract Administrator needs to prepare a thorough description of the work required for an order, the specification clauses serve to prevent the scope, standard and cost of work escalating. The comprehensiveness and stand-alone nature of the specification clauses are factors that make pre-priced standard Schedules of Rates ideal for use as tendering and contract documents.

Tender assessment – The Schedule of Rates allows a comparison between Contractors based on price. This comparison can be very straightforward if the tender requires a single tendered percentage for all work. It is slightly more complicated if the tender requires separate percentages for different trades, locations, value of Works Orders, etc. Each percentage will need to be weighted according to frequency of occurrence (see Chapter 5 for more on tender evaluation). However, every rate will be different if a blank Schedule has been issued for pricing by Contractors and, consequently, assessment would be extremely time consuming.

Basis of payment – The priced Schedule of Rates and its attendant percentage additions or deductions, allow for work to be carried over the period of the contract at an agreed price, even though the exact composition of the work and its timing are not known at the time of tendering. This use of the Schedule is the basis of Measured Term Contracting.

Subsidiary functions

Budget Planning – The priced Schedule of Rates enables managers to estimate with a degree of accuracy the cost of their programme and to compare alternative programmes at an early stage.

Cost control of the budget – The ability to value the Works Orders as they are issued at agreed rates allows managers to know the amount of work committed and therefore to control the flow of work to meet budget requirements.

Ordering of work – Certain types of Schedules of Rates can form the basis of a Works ordering system. For example, a Composite Schedule of Rates may contain descriptions of complete tasks such as 'Forming a new opening in a wall, providing a new window, glazing, plastering the reveals and painting'. Such a description could be referred to on the Works Order form by its 'Item number' in the Schedule. This simplifies administration as the invoice will exactly match the Order and reduce the need for measurement and checking.

Contents of a Schedule of Rates

The exact contents of the Schedule of Rates will vary according to requirements of the individual contract, but Schedules would normally contain preliminaries, preambles and priced items of work.

Preliminaries

The preliminaries should define the scope of the contract, detail the conditions of contract and clarify any matter that the Employer wishes to bring to the Contractor's attention. Where a standard published Schedule of Rates is used, the preliminaries may take the form of a separate document.

Where a standard form of contract is used, the preliminaries only need to contain the clause headings and additional special conditions. If the Employer's own procurement contract is being used, the preliminaries may well contain the detailed contract conditions.

Typical matters that may need to be covered include:

- scope of the contract – including name of the Employer and supervising officer
- provision for determination of the contract by either party
- method of ordering work and the time for completion
- liquidated damages for delayed completion
- compliance with statutory regulations

- plant, tools and vehicles; provision and maintenance of scaffolding
- responsibilities for the provision of materials including materials supplied by the Employer
- consent required for subletting of the contract
- access to the site or property by the Contractor – including provision to enable the Contractor to carry out the work
- valuation of the work – in accordance with the Schedule of Rates and the percentage adjustment
- agreement of rates where prices in the Schedule of Rates may not apply – including provision for dayworks
- interim payments on monthly valuation
- certification for payment by the supervising officer of the Contractor's account within a specified period
- defects liability – provision for the Contractor to make good any defects appearing within a specified time after completion of the work
- materials, works and workmanship to conform to description
- removal of rubbish
- insurance of the works against fire, etc., and insurance of existing structures
- insurance against damage to property and injury to persons – third-party insurance
- assignment of the contract
- Value Added Tax
- methods of settling disputes
- contract definitions – the Contract, Employer, Contractor, Contract Area, Contract period, Works Order or Orders, Site, Schedule of Rates, Contract Administrator/Project Manager, Percentage Adjustments.

Most of these items are covered by Standard Forms (see Chapter 9).

Preambles

Sufficient specifications preambles should be included to define the materials and workmanship required. They should not be over elaborate and should reflect the nature of maintenance work. There is sometimes a tendency to pad out the preambles with masses of specifications for new work when the Schedule of Rates only has items for repairs. For example, there is no point reproducing pages of new work plastering specifications if all that is included in the Schedule is patch repairs to plaster.

Priced rates for items of work

The Schedule of Rates should contain the descriptions of the work and the price to be paid, subject to tendered percentages. Some approximation of the annual frequency of each item could be included. This information is rarely available, but it would undoubtedly be useful to Contractors if it were.

It is normal to issue a Schedule of Rates with all inclusive prices but, in the past, Measured Term Contracts have been run using Schedules with labour-only rates, leaving materials costs to be reimbursed at cost.

There have been some instances of blank Schedules of Rates being issued at tender stage for Contractors to price. This requires a lot of work by the tenderers and can lead to some problems at the tender assessment stage, as every rate for every tenderer will be different. This makes pricing blank Schedules fairly risky for both the Contractor and the Employer.

Attributes of a good Schedule of Rates

A Schedule of Rates should be:

- concise
- complete
- clear
- consistent.

Concise

In content – The Schedule should not contain items that will never be used. The inclusion of unnecessary items will mislead the Contractor as to the nature of the work and make the document unwieldy in use. It is pointless to include rates for new work or, for example, areas of roof tiling if the work to be ordered will only ever be the replacement of tiles in ones and twos.

In presentation – The Schedule should avoid repetition by use of headings for details which are common to a succession of jobs, and by not attempting to differentiate between types of work which are not significantly different in nature or price.

Complete

The Schedule should cover all the types of work that need to be carried out under the contract. It is impossible to avoid some Dayworks on a maintenance contract, but the drafting of the Schedule should endeavour to keep this to a minimum.

There obviously has to be some trade-off between conciseness and completeness. However, a working knowledge of the estate to be maintained should indicate the scope of the items to be included. The 'we will stick it in just in case' approach should be avoided.

If the scope of the Schedule does reflect the work that is to be carried out, it will help the Contractor in understanding the nature of the work for which he is tendering and should lead to better value for money for the Employer.

Clear

The items should be unambiguous as to specification, shape, size, etc. The descriptions should define the work included so that arguments do not arise during the contract. The unit of measurement should also be unambiguous. The use of a Standard Method of Measurement will assist definition.

Consistent

The pricing level should be consistent throughout the Schedule. It makes it difficult for Contractors to tender a single percentage on a Schedule if the rates within each trade, or even between trades, are unevenly priced. No two surveyors or Contractors will price a selection of items the same. What may seem to be consistent pricing to one person will be considered uneven to another, but obvious inconsistencies should be avoided.

The pricing should also be a reasonable market price. In theory, the Contractor is pricing independently of the Schedule but there must be some doubts raised if the rates presented require a large adjustment. Low rates requiring a big addition may well make the Contractor suspicious of the whole exercise and while Contractors will and do tender negative percentages, there is probably a psychological barrier to putting too large a negative percentage in a tender.

Types of Schedules of Rates

An ideal Schedule of Rates would be one that contains the exact description of the work to be carried out on every Works Order. Given the range of possible items and the possible combinations of items, the permutations involved are probably infinite. There are two possible solutions to this problem, which define the two main types of Schedule of Rates available:

1. Detailed Schedules
2. Composite Schedules.

The Detailed Schedule of Rates contains individual items of work for separate tasks. This allows a job to be priced by building up from individual items of work, very much like a traditional estimating price book.

The Composite Schedule of Rates contains tasks built up from the individual items of work involved in carrying out the task.

For example, if the actual work involves replacing a handbasin – including replacing the wall tiles in the splashback, resealing the basin and overhauling the taps – the Detailed Schedule will have separate items for replacing the basin, replacing the tiles, sealing the basin and overhauling the taps. The Composite Schedule may have a single item for replacing a hand basin including all making good to tiles and plaster, overhauling taps and renewing back nuts and traps where necessary.

There are advantages and disadvantages to both types of Schedule of Rates.

Detailed Schedule of Rates

Advantages:
- There is greater accuracy of payment as the value of the Order is for work actually carried out and, therefore, there is less risk to the Contractor. This may possibly produce keener pricing.
- There is increased flexibility as the larger number of individual items of work allows for variations in the Work Order and the Schedule of Rates is more likely to be applicable in different situations.

Disadvantages:
- There is a greater requirement for measurement and valuation and this will increase administration and, possibly, overall costs.
- There is less accuracy in estimated values of Works Orders. These will be 'guesstimates' unless pre-measurement is carried out.
- They are not suitable for providing the basis of Works Orders without pre-inspection.

Composite Schedule of Rates

Advantages:
- There are reduced administration costs as there is less requirement for pre- or post-measurement. Descriptions of tasks on the Works Orders are more likely to be the same as those provided in the Schedule.

- They are quicker and more economical in situations where Works Order values are low. Building up low-value Works Orders from several individual items is not cost-effective.

- There is greater certainty of committed money as the Works Order estimated value and final value will be similar or even identical.

- The composite descriptions can be used for placing Work Orders. This reduces the need for an exact diagnosis of faults by the person issuing the Order as the repair item will cover a variety of possible tasks required to solve a problem. It is, therefore, not necessary to predetermine which task will be required when ordering the work.

- The Composite Schedule will help to define the work. The Schedule will indicate the scope of work likely to be included in the contract.

Disadvantages:
- It lacks flexibility in that it has limited use if work beyond the original scope of the contract arises. Furthermore, the Schedule is not portable in that it cannot be used on a different estate or by a different Employer without significant amendment.

- There is a pricing risk in that the pricing requires some averaging of the likely work to be carried out and the resultant rates will rely on 'swings and roundabouts'. This increases the risk to the Contractor, which will be reflected in his tender and also introduces some risk to the Employer.

Choice of Schedule of Rates

There is a trade-off between increased accuracy of payment and increased cost of measurement and administration in choosing between these two types of approach.

The two main factors affecting the choice are the nature of the estate and the likely value of Works Orders. If the estate contains a fairly narrow band of building types with similar recurring maintenance problems, a Composite Schedule of Rates would seem more appropriate. If the estate to be maintained is likely to produce a wide variety of different items of work, because of the nature of the buildings or the variation in their specification or use, then a Detailed Schedule of Rates is more appropriate. Housing Schedules of Rates should be as simple as possible providing identification of complete jobs rather than items of work.

If the average value of the Works Orders is small, the cost of measurement or checking becomes prohibitive. However, if the average Works Order value is large, particularly if it is likely to contain more than one job, the post-measurement becomes a requirement and the increased accuracy obtained from a Detailed Schedule of Rates probably outweighs the administration costs.

Table 8.1 below shows the factors that would tend to make one type of Schedule of Rates more or less applicable.

Table 8.1   Choice of Schedule of Rates

Detailed Schedule	Composite Schedule
High average Works Order value	Low average Works Order value
Disparate building types	Uniform building types
Varied specification	Consistent specification
Varied usage of building	Consistent usage of building

It is possible to mix the two approaches to provide composite rates for regularly occurring items, and detailed rates for less predetermined work. However, care must be taken in mixing detailed and composite rates to ensure that the composite items are clear in their intent and do not encourage the Contractor to add detailed rates for work that should be covered in the composite item. Nor should the Contractor be left to choose whether to use the composite item or the detailed measurement.

Availability of Schedules of Rates

In setting up a Measured Term Contract there are two broad options for acquiring an appropriate Schedule of Rates:

1. use a published Schedule of Rates
2. have a bespoke Schedule of Rates tailor-made.

There are advantages and disadvantages with each solution.

Published Schedule of Rates

Advantages:

- Familiarity – many Contractors' consultants and Employers will be familiar with the Schedule, which could produce more confidence among tenderers and consequently keener tenders.
- Availability – they are readily available. They can be obtained from specialist bookshops or direct from the compilers.
- Established Practices – they may have existing procedures developed for operating term contracts, or there may be computer programmes available to use with the Schedules.
- Format – they tend to be detailed rather than composite, which is an advantage if a Detailed Schedule is required.
- Cost – relatively cheap for a control document for fairly large contracts.

It is worth bearing in mind that a Bill of Quantities for a £1,000,000 new work contract will cost something in the order of £10,000 to £15,000; a Measured Term Contract with a similar volume of expenditure should be regarded as in need of equally valuable documentation. The cost of a published Schedule of Rates and associated administrative documentation are small compared with the benefits.

Disadvantages:

- Loose-fit – they will not be concise and will contain items that are not appropriate to a particular estate which, in turn, may make them unwieldy.
- Non-specific – they do not help to define the type of work to be carried out by the Contractor.
- Format – they tend to be detailed rather than composite, which is a disadvantage if you want a Composite Schedule.
- Compatibility – may lack compatibility with existing procedures and programmes.

Bespoke Schedules of Rates

Advantages:

- Concise – can be concise and specific to the estate and therefore easier to use.
- Format – can be detailed or composite as appropriate.
- Specific – help to define the work to be carried out by the Contractor.
- Ordering of work – they can be designed to facilitate the placing of Works Orders.

- Compatibility – they can be designed to fit with existing procedures and computer systems.

Disadvantages:
- Unfamiliarity – there will be a learning curve for both the Contractor and the Employer.
- Cost of production – the cost will obviously vary depending on the range of work and number of rates, but something similar to the cost of preparing a BQ will give an indication. Whatever the cost, it will be negligible compared with the amount of money spent using it.
- Availability – it will take some time to prepare.

There is, of course, an option to mix the two, using the published Schedule of Rates together with a bespoke Schedule of Rates for more particular items of work.

Published Schedules of Rates available

PSA Schedule of Rates

These are published approximately every five years. The Schedules are comprehensive, cover new work as well as maintenance and are clear and detailed in their specification. The presentation tends to be tabular and is available in an electronic form. A bespoke service is offered whereby Employers can have a Schedule of Rates prepared based on their particular requirements.

Since the privatisation of the PSA, the Schedules of Rates have been produced by Carillion Services and published by The Stationery Office. The current suite of commercially available Schedules comprise:

- *Building Works* – this covers all aspects of general building work. It includes approximately 21,000 work items/rates.
- *Mechanical Services* – this provides comprehensive coverage of all aspects of mechanical services. It includes approximately 11,000 work items/rates.
- *Electrical Services* – this is a companion volume to the Mechanical Services Schedule, and covers all aspects of electrical services. It includes approximately 10,000 work items/rates.
- *Decoration Works* – this is applicable to complete packages of decoration and redecoration works.
- *Landscape Management* – this covers all aspects of grounds maintenance, hard and soft landscaping and street furniture.
- *Road Works* – this was produced in compliance with the Department of Transport Highway Works documentation and covers work from footpaths to trunk roads.

The rates are calculated from first principles and take account of labour constants derived from research and historic data, industry-accepted labour rates, materials costs derived from manufacturers' costs/suppliers' price lists (adjusted for trade-discounts as appropriate) and plant costs derived from research and historic data. The rates also include an allowance for preliminaries (site overheads). The rates are all expressed at the price levels applicable to the base-date of the Schedule. The base-date is the reference point used for updating the rates in the Schedule.

The Department of Trade and Industry (DTI) independently produces and publishes percentages specifically for the purpose of updating the PSA Schedules of Rates.

National Schedule of Rates

Introduced in the early 1980's as a joint venture between the Society of Chief Quantity Surveyors in Local Government and the Building Employers' Confederation, the Schedule covers all normal building trades and new work as well as repairs and alterations. The format is much like a Bill of Quantities but includes separate material, labour and plant costs.

The National Schedule of Rates also produce Mechanical and Electrical Services and a separate Painting and Decorating Schedule of Rates. There is also a separate document giving the labour constants used in the Schedule. The Schedule is repriced annually.

National Housing Federation (NHF) Schedule of Rates

Introduced in the 1990s by the National Housing Federation, the Schedule is widely used for repairs and maintenance to residential dwellings across all housing sectors. It is a fully priced, composite schedule, complete with contract documentation for letting a Measured Term Contract for day-to-day and void repairs.

The last major re-pricing of the schedule since it was developed was issued in 1999 at January 1999 prices.

The Building Maintenance Information (BMI) Price Book

The BMI Price Book was prepared originally as an estimating guide for maintenance work but it has been used quite widely as the basis for Measured Term Contracts. It is produced in an A5, ring-bound format and is available in an electronic form. BMI also offer a bespoke service whereby Employers can have a Schedule of Rates prepared based on their particular requirements. Although concentrating on repair and maintenance work, it also contains an element of minor new works. All items have details of labour times and labour- and material-costs and are, for the most part, expressed as composite rates.

The Price Book is re-priced in its entirety each year; in addition, the rates within the book can be updated by using BMI Maintenance Cost Indices which are calculated and published both monthly and quarterly. Both the book and the indices are published by Building Cost Information Service Ltd, a company of the Royal Institution of Chartered Surveyors.

Fluctuations

Whichever Schedule of Rates is chosen, there will be a requirement to allow for increased costs where contracts are let over a period longer than a year. The major published Schedules of Rates allow for this.

The PSA Schedule of Rates use the application of 'Updating Percentage Adjustments for Measured Term Contracts' on a monthly basis. These percentages, which are issued monthly by the Department of Trade and Industry (DTI), provide a simple way of bringing the Schedule of Rates up to current price levels, and are generally accepted by the building industry. They are produced monthly, by applying the increased cost of labour, material and plant to a statistical sample of items in the Schedule of Rates. The sample of items is re-assessed when a new Schedule is produced.

The National Schedules of Rates provide for fluctuations by being re-issued each year, whilst the BMI Price Book allows adjustment through use of indices and by being re-issued each year.

For bespoke Schedules of Rates, one of the existing indices issued by BMI or DTI could be used. However, it might be more appropriate to compile an index based on the types of material and labour actually being used in the contract. A fairly simple model for an index can be compiled from the available estimating information and the frequency of ordering various types of work. The weightings and proposed source of updating should be included in the contract documents. It is advisable to use published sources for updating the indices in order to avoid disputes.

A specially compiled index has the advantage that if, for example, a particular type of kitchen unit is always specified, it is fairer to reflect changes in their price rather than general trends in joinery prices.

9 Standard Forms of Contract

Introduction

Standard Forms are produced through wide consultation with the industry as a whole, leading to general acceptance of the final conditions. It is advisable to use a Standard Form wherever possible as they are much more familiar to users, easier to use (Model Forms, etc.) and much less likely to generate costly disputes.

There are two major Standard Conditions of Contract suitable for Measured Term Contracts:

1. the JCT Standard Form of Measured Term Contract 1998
2. GC/Works/7 (1999) Measured Term Contract (copies of GC/Works/7 Model Forms are featured in this book's Appendix 1).

The information required from the employer

In addition to the Standard Conditions, there are various pieces of information that should be provided at tender stage. These include:

Properties in the contract area

The properties to be maintained should be clearly defined. The Contract Area itself is to be stated in the recitals to the Articles of Agreement and Item 1(a) of the Appendix in the JCT Standard Form, and in the Abstract of Particulars (see Model Form 1) in GC/Works/7.

Type of work to be carried out

There should be a clear description of the types of work for which Orders may be issued under the Contract. This should include such information as whether it is normal maintenance, decoration, etc. (Item 1(b) of the Appendix to the JCT Standard Form).

Order values

The minimum and maximum value of any single Order to be issued under the Contract should be defined. This is to be stated in Item 2 of the Appendix in the JCT Standard Form, and in the Abstract of Particulars (see Model Form 1) in GC/Works/7.

Approximate value of the work to be carried out

Approximate value of the work to be carried out should be stated at Item 3 of the Appendix in the JCT Standard Form. The JCT Standard Form makes it clear that this is an approximate value and that if it proves to be incorrect there are no penalties on either side. It is intended as a guide to the Contractor. There is no provision for this in GC/Works/7 so separate provision should be made elsewhere in the contract documents.

The contract period

It is suggested that this should never be less than a year and is normally three years. This is to be stated in Item 4 of the Appendix in the JCT Standard Form, and in the Abstract of Particulars (see Model Form 1 in GC/Works/7).

Priority coding for orders

This enables the employer to request different response times from the Contractor. For example, he may require an Order to be carried out within four hours if a roof is leaking, while 24 hours might be more appropriate for other repairs. This is to be stated in Item 4 of the Appendix in the JCT Standard Form. There is no provision for this in GC/Works/7 so separate provision should be made elsewhere in the contract documents.

The following is a typical example of priority categories: emergency work, immediate work, urgent work and non-urgent work. It is intended as a guide as the JCT Form refers to priority work, but does not give examples.

Classification of urgent and non-urgent work
Category 1: Emergency work – to be done within four hours of being reported.

These should be restricted to repairs that, in the opinion of the Employer's representative, would be likely to cause danger to health or life and/or which could directly affect the safety of the building. More specifically, such repairs may include, but are not limited to:

- gas leaks
- electrical faults meeting the criteria mentioned above
- water leaks and burst pipes where the tenant is unable to isolate the problem
- blocked drains and toilet pans, before 21:00 hours
- jammed entrance doors and loss of keys where no alternative access is available (flats and upper maisonettes)
- lightning, flood, fire and storm damage, etc.
- structural danger to tenants and general public
- assisting in re-accommodating occupiers and making secure fire, flood and storms, etc. damaged properties
- making dwellings secure.

Category 2: Immediate work

These should be restricted to repairs requiring attention within 24 hours if determined to be a matter for immediate action, taking into account the nature of the repair. Such repairs may include, but are not limited to:

- central-heating repairs
- lighting and power-failure
- blocked drains and toilet pans after 21:00 hours.

Category 3: Urgent work

These should be restricted to cases where leaving the work more than ten days would cause further non-severe damage to the property, or greatly affect the use of the property and cause discomfort to the tenant. Such repairs may include, but are not limited to:

- roof leaks (where water penetration can be contained by tenant)

- replacement of cracked and leaking/broken sanitary ware (excluding WC)
- repairs to central-heating system (in winter months)/domestic hot water systems
- defective joinery items likely to have a serious effect on the usage of the property, e.g., window sash, handrail, staircase
- defective or jammed external door locks
- serious structural defects, e.g., badly cracked/defective ceiling
- follow-up repairs to emergency/immediate work
- guttering in winter months
- leaking overflow
- defective water-heater.

Category 4: Non-urgent work

1. Work to be completed within one month. These may include, but are not limited to:

- non-programmed joinery repairs where the component is not fulfilling its function
- replace cracked and leaking/broken 2nd WC
- replacement of white glazed sinks where considered to be unsanitary (excludes cracked/broken sinks).

2. Work to be completed within three months. These may include, but are not limited to:

- defective principal access paths
- structural defects to boundary walls
- repointing in isolated areas of the brickwork
- rebuilding of chimney (where not immediately dangerous)
- all other repairs considered to be necessary and not included in other categories or programmed.

Schedule of Rates

The JCT Standard Form was originally designed around the National Schedule of Rates. However, much that was specific to the National Schedule has been removed from this document so that it is now usable with the PSA Schedules of Rates or with other individual Schedules. Item 6 of the Appendix in the JCT Standard Form gives the National Schedule of Rates at (a) and allows an alternative Schedule to be identified at (b). The Tender Form (see Model Form 3) in GC/Works/7 allows the insertion of the Schedule of Rates at Item 1(e).

Fluctuations provision

Details of any method of allowing for fluctuations should be given. The contract can be fixed price (see Item 8 of the Appendix in the JCT Standard Form). Fluctuations can be allowed for by application of an index on a monthly basis as with GC/Works/7 (see Item 1(f) of the Tender Form – Updating Percentage Adjustments for Measured Term Contracts). Alternatively, they can be allowed for by annual re-pricing of the base Schedule as with the National Schedule of Rates (see Item 7 of the Appendix in the JCT Standard Form) or any other method specified at Item 7 of the Appendix.

Method of calculating Daywork

This allows work to be done on a Daywork basis. It is important to make sure that the basis on which the Daywork is to be paid is appropriate and clear.

Item 9 of the Appendix to the JCT Standard Form gives the option of percentage adjustments to the definition of the prime cost of Daywork or the insertion of inclusive hourly rates. If the Employer has the expertise to make judgements on the inclusive hourly rate offered, there may be an advantage to him in adopting this latter option. GC/Works/7 offers the choice of definition on the Tender Form (Model Form 3) at Item 1(g) and provides a comprehensive Contract Daywork Schedule (Model Form 4).

Responsibility for Measurement

Measurement details of Orders will be taken by the Employer's representative or made under the provisions for self-measurement by the Contractor. This is to be stated in Item 10(a) of the Appendix in the JCT Standard Form. In GC/Works/7, the Contractor is responsible for measurement although the Quantity Surveyor has the right to be present and the Project Manager has the power to instruct joint measurement (see Condition 26(1)).

Progress payments

The estimated value of orders above which progress payments will be allowed should be defined. This is to be stated in Item 10(b) of the Appendix in the JCT Standard Form with a default value of £1,500, and in the Abstract of Particulars (see Model Form 1) in GC/Works/7 (see the notes against Condition 29).

Requirements for Health and Safety

Article 4 of the JCT Standard Form identifies 'the Planning Supervisor' and Article 5 identifies 'the Principal Contractor' as required by the Construction (Design and Management) Regulations 1994. The JCT Standard Form requires the Contractor's policy statement to be submitted with his tender (see Item 11 of the Appendix in the JCT Standard Form). GC/Works/7 provides for an Outline Health and Safety Plan to be included with the Invitation to Tender (Model Form 2, Item 1(d)).

Insurance

Details of the level of insurance required under the contract. In the JCT Standard Form, Appendix, Item 13, the insurance cover (to be taken out by the Contractor (Clause 6.4.1)), required for any one occurrence or series of occurrences arising out of one event (Clause 6.4.2) and the percentage to cover professional fees (Clause 6.9) incurred by the Employer are both required to be filled out, as is the Contractor's annual policy renewal date. In accordance with existing Government policy decisions, there is no provision for insurance in GC/Works/7.

Break provision

This is the period of notice to be given by either party to determine the employment of the Contractor. It is fundamental to the Contract that there is a good relationship between the Employer and the Contractor. The 'break' provision means that if the contract is not working out for either party, the Contract can be determined. For example, if the Contractor is not responding satisfactorily or if the amount of work or its nature is vastly different from that indicated in the Contract, the 'break' provision offers a way out without protracted disputes. Notice can be given by either side after a pre-determined period with no penalties. In the JCT Standard Form, the default is 13 weeks (Clause 8.1), but there is provision for a lesser period in the Appendix, Item 14. GC/Works/7 specifies three months' notice (Condition 7.2).

Settlement of disputes

Details of how disputes are to be settled should be defined. The JCT Standard Form, Appendix, Item 15, provides for the nominating body responsible for appointing an Adjudicator to be stated. The Abstract of Particulars (Model Form 1) of GC/Works/7 makes similar provision (see the notes against Condition 39). Model Form 5 provides an Adjudication Agreement. The JCT Standard Form, Item 16, also provides for the option of Arbitration or Legal Proceedings.

Details to be inserted by the Contractor

The Contractor is required to complete the Contract by providing his tendered terms of payment and other information as follows:

Percentage addition or deduction to the stated Schedule of Rates

Contracts may allow for a single percentage, but there is nothing to prevent the Employer from requesting differing percentages to cover a variety of circumstances. For example, size of Order, different locations, different categories of work, or for separate sections of the Schedule of Rates. Item 6 of the Appendix to the JCT Standard Form provides for one percentage (Percentage 'A'). However, the Tender Form (see Model Form 3) to GC/Works/7 provides for three percentages to be applied to value bands £0–£5,000, £5001–£25,000 and exceeding £25,001, at Item 5(1).

Daywork provision

Item 9 of the Appendix to the JCT Standard Form sets down the various percentage adjustments to the definition of the prime cost of Daywork for overheads and profit on labour, materials and plant. There is also the option to insert inclusive hourly rates. GC/Works/7 provides a Contract Daywork Schedule for hourly rates (Model Form 4) for the tenderer to sign.

Non-productive Overtime

Contractors may be asked to quote separately a percentage on the net cost of overtime. When the Contractor carries out work that will be valued using the Schedule of Rates, but some, or all, of the work will be executed outside normal working hours, the Contractor is entitled to recover the extra element of overtime he has incurred. This becomes payable only when overtime working is instructed (see Item 9(f) of the Appendix to the JCT Standard Form). GC/Works/7 gives no option but states that non-productive time is reimbursed as normal Daywork multiplied by 0.5 being the enhancement of time laid down in the Working Rule Agreement (see Model Form 4, Item 7(4)).

Contractor's safety policy

The majority of Orders raised under a Measured Term Contract will not be affected by the Construction (Design and Management) Regulations 1994 (SI 1994 No. 3140), but there are provisions in the JCT Standard Form and GC/Works/7. Article 4 of the JCT Standard Form identifies 'the Planning Supervisor' and Article 5 identifies 'the Principal Contractor' as required by the CDM Regulations. The JCT Standard Form requires the Contractor's policy statement to be submitted with his tender. GC/Works/7 provides for a statement by the Contractor on the Tender Form (Model Form 3) confirming that he will provide a statement and details of how he will implement and develop the Outline Health and Safety Plan included with the Invitation to Tender (Model Form 2).

Insurance

The Contractor must maintain a Joint Names Policy for All Risks Insurance and the annual renewal date must be stated (see Item 13 of the Appendix to the JCT Standard Form).

JCT Standard Form of Measured Term Contract

The Joint Contracts Tribunal (JCT) Standard Form of Measured Term Contract was first published in 1989 as a result of a request from local authority associations for a Contract to meet growing demand in this field. It was based on a contract that the Society of Chief Quantity Surveyors in Local Authorities (SCALA) and the Building Employers' Confederation (BEC) produced in 1982. There had been a move away from Directly Employed Labour (DEL) towards 'contracting out' and it was apparent that local authorities were not to be allowed to do as much maintenance work in-house.

The JCT Drafting Sub-Committee is responsible for drafting all the contracts, sub-contracts, practice notes and amendments. All the constituent bodies of the JCT are consulted and, if none object, a working party is appointed to prepare the draft of the contract. The working party consults experts in the particular topic, in this case people who had knowledge of Measured Term Contracts, so that practical experience is reflected in the draft.

The JCT have also produced Practice Note MTC/1 and Guide to the Standard Form of Measured Term Contract' (JCT, 1989), a useful document for those using the Contract Form.

The following is a brief summary of the Contract's Clauses but should not be regarded as a substitute for reading the Contract or Practice Note.

Articles of Agreement

The parties to the contract, 'the Employer' and 'the Contractor', are identified and, in the 1st Recital, 'the Contract Area' is defined. The 1st Recital also requires Maintenance and Minor Works to be carried out in accordance with the details set out in the Appendix, Items 1 to 5. The 2nd Recital states that the Contractor has offered to carry out the works upon the Terms of Payment recorded in the Appendix, Items 6 to 10.

Article 1 states that the Contractor will carry out the works and Article 2 states that the Employer will pay the Contractor, subject to the Conditions.

Article 3 identifies 'the Contract Administrator' who is responsible for administrating the Contract on behalf of the Employer. This is an important appointment as he will be responsible for issuing most instructions to the Contractor. The person who fulfils this role may vary from an architect, engineer or building surveyor, to someone who has been promoted from being 'on the tools', but experience of this specialised type of contract is the most obvious advantage. The Employer's role is to make payments and prevent obstructions to the Contractor in carrying out the ordered work. It should be noted that, unlike GC/Works/7, there is no 'Quantity Surveyor' named in this Contract, such functions being carried out by the Contract Administrator (see Section 4).

Article 4 identifies 'the Planning Supervisor' and Article 5 identifies 'the Principal Contractor' as required by the Construction Design and Management (CDM) Regulations.

Articles 6, 7A and 7B identify the method of settling disputes or differences. Article 6 allows either party to refer a dispute to adjudication in accordance with Clause 9A. If the entry in the Appendix, Item 16, states 'Clause 9B applies', Article 7A permits, in specified circumstances, disputes to be referred to arbitration. If the entry in the Appendix, Item 16, stating 'Clause 9B applies' is deleted, then disputes are to be determined by legal proceedings.

It is a difficult pre-Contract judgement to be made when no dispute has yet materialised (and, therefore, cannot be categorised), as to which of the above three options will provide the Employer, and/or the Contractor, with the most satisfactory option in respect of dispute resolution. If speed of resolution of a dispute is the main criteria for the Employer, it may be that, when filling out the Appendix, Adjudication will be the most suitable option.

Section 1   Intentions of the Parties

Clause 1.1 (Definitions, etc.)
Paragraph 1 gives the definitions of terms to be used throughout the Contract (not just the Conditions). Most of the definitions are self-explanatory. Some, such as 'the Contract Area' and the 'Contract Period', are cross-referenced with the Recitals or Appendix. It is essential that such information be inserted if the Contractor is to be expected to submit a reasonable tender.

Clause 1.2 to 1.5 (Contractor's obligations)
Clause 1.2.1 describes the Contractor's obligations, in general, to carry out work in a good and workmanlike manner. In Clause 1.2.2, it also reserves the right of the Employer to have work in the Contract Area carried out by other Contractors. By entering into a Measured Term Contract, the Contractor is not absolutely entitled to all the work in an area, or on particular buildings, carried out under his Contract.

The Contractor must, subject to Clause 1.6.1, provide all the resources necessary for carrying out any Order and to comply with his statutory duties in respect of safety, health and welfare.

Clause 1.6 to 1.11 (Right of Employer to supply materials or plant)
Under these Clauses, the Employer is entitled to supply materials or plant to the Contractor. The materials remain the property of the Employer and cannot be used by the Contractor for any purpose other than the carrying out of an Order. The Contractor is responsible for the safe storage of materials and plant, the disposal of any materials not required and the return of any Employer-supplied plant or equipment.

The Contractor is entitled to payment for disposal and for a handling charge of 5% of the value of Employer-supplied materials.

Clause 1.12 (Materials – quality)
The quality of materials is specified in the Schedule of Rates referred to in the Appendix, Item 5, but Clause 1.2.1 (Contractor's obligations) reserves the right of the Contract Administrator (CA) to approve the quality of materials.

Clause 1.13 (Value of work to be carried out under this contract)
The Appendix, Item 3, makes provision for an approximate anticipated value of work to be carried out under the Contract to be inserted. Clause 1.13, however, states that the Employer 'accepts no liability as to the actual amount of work that will be ordered'.

This is a very important provision and is the subject of frequent disputes on Measured Term Contracts. Although the value of work to be ordered is not guaranteed, every effort should be made to ensure that this amount is reasonably accurate, if a good working relationship between the parties is to be maintained. The Contractor will have provided resources appropriate to the value of work and will not be content to have workpeople standing idle.

Clause 1.14 (Schedule of Rates – relation to Articles, Conditions and Appendix)
This Clause gives the priority of documents contained in the Contract. In addition to the Schedule of Rates, it refers to drawings and specifications. The Conditions themselves do not contain any specifications, or preliminaries for that matter. These can be incorporated in the Schedule of Rates or they can be issued as a separate document. Preliminaries particular to the circumstances of the Contract may be incorporated. The term 'Schedule of Rates' is used in the conditions to embrace preliminaries and specifications as well as the priced Schedule of Rates.

Clause 1.15 (Programme)
This Clause gives the Contract Administrator the right to request a programme from the Contractor for Orders identified by the Contract Administrator in his request. However, any programmes requested do not impose any obligations beyond those in the Contract Documents.

A well-advised Employer would specify that he requires the supply of the programme in the form of a particular software disk, as well as hard, paper copies. This is so that the Employer can interrogate the programme in order to ascertain what parameters have been used to determine it. For example, the Employer should be able to ascertain that the supplied programme is resource- (and not data-) driven, and the quantum of those resources. This information is very valuable to the Employer and cannot be obtained from a hard copy.

Clause 1.16 (Reappointment of Planning Supervisor or Principal Contractor – notification to Contractor)
The Articles name the Planning Supervisor and the Principle Contractor in accordance with the CDM Regulations. This Clause requires the Employer to notify the Contractor of any change in appointees.

Clause 1.17 (Giving or service of notices or other documents)
This Clause describes the circumstances under which giving or serving notice to the Contractor is treated as having been effectively served.

Clause 1.18 (Reckoning periods of days)
The Contract requires various acts to be done within specified periods. For example, Clause 4.15 states that certificates must be paid by the Employer within 14 days of issue by the Contract Administrator. Clause 18 states that a specified period from a date starts immediately after that date and excludes Public Holidays.

Clause 1.19 (Applicable law)
The law applicable, wherever the Works are situated, is the law of England, although a Footnote gives the parties the opportunity to amend this Clause.

Clause 1.20 (Contracts (Rights of Third Parties) Act 1999 – contracting out)
This confirms that the Contract does not confer any right to enforce any of its terms on any person who is not a party to it.

Section 2   Commencement and completion

Clause 2.1 (Orders to be executed within the Contract Period)
All orders must be in writing (see Clause 1.1, Definitions), which can include drawings and variations. If the Employer's organisation cannot comply with this stipulation, he must amend the Contract. Orders must be reasonable and capable of being carried out within the Contract Period.

Clause 2.2 (Orders – completion)
This Clause describes the information the Order must contain, including the date of commencement and reasonable date for completion of the Order, subject to any priority coding. The Appendix, Item 4, makes provision for the classification of orders with a different priority. There is no provision for different percentage adjustments to the Schedule of Rates to be applied to these types of work. However, it is possible, when seeking tenders, to allow the Contractor to put a premium on orders to which he has to respond very quickly. This could be done adding a schedule similar to the following:

(a) for orders for which the total of all the **net updated value** of measured work, in all sections of the Schedule of Rates, falls within the Priority Code Category 1

+Add/–Deduct
...................%

(b) for orders for which the total of all the **net updated value** of measured work, in all sections of the Schedule of Rates, falls within the Priority Code Category 2

+Add/–Deduct
...................%

(c) for orders for which the total of all the **net updated value** of measured work, in all sections of the Schedule of Rates, falls within Priority Code Category 3

+Add/–Deduct
...................%

Clause 2.3 (Matters causing delay – fixing later date for completion)
The Contractor must give notice of any matter which is likely to cause, or is causing, delay in completing an Order. Whether or not he gives notice, the Contract Administrator has to give a revised completion date if the circumstances are beyond the control of the Contractor.

If the Contractor is unable to carry out an Order using 'his best endeavours', Clause 3.10 enables the Contract Administrator to arrange, after the giving of notice, for a third party to carry out the work and recover the cost and any reasonable consequent expenses from the Contractor.

Clause 2.4 (Defects)
There is a six-month defects liability period following the completion of each Order. There is no retention of money in respect of any defects found. On a Term Contract, it is in the Contractor's interests to rectify defects. The Employer has the right to deduct money from sums due for other works to carry out any remedial work not done by the Contractor.

There are no liquidated or ascertained damages in the Conditions. If the Contractor is late and the Employer suffers loss, he may be able to recover proven loss at common law.

Section 3   Control of work or supply comprised in Orders

Clause 3.1 (Assignment)
This Condition prohibits either the Contractor or the Employer from assigning or transferring the Contract without the other's consent. This is different to the Government Contract GC/Works/7, which permits the Employer to assign without the Contractor's consent.

Clause 3.2 (Sub-contracting by Contractor)

The Contractor is permitted to sub-contract subject to the consent of the Contract Administrator. As with most JCT Contracts, this contract relies on the Contract Administrator, who is named in Article 3, to administer the Contract on behalf of the Employer.

Clause 3.3 (Contractor's representative)

This Clause requires a competent and accessible representative to be employed by the Contractor to receive Orders or Variations from the Contract Administrator. It does not specify that the agent is to be site-based, but on a Contract of a significant size this may be desirable.

Clause 3.4 to 3.5 (Access to the site)

Clause 3.4 states that, unless it is otherwise provided for by the Schedule of Rates, the Contract Administrator is responsible for arranging access to the Site. If access is not available, the Contractor is permitted to recover additional costs for the unproductive time that he incurs, on a Daywork basis.

In practice, Employers frequently prefer the Contractor to arrange for his own access, particularly when working in occupied premises. Other Contracts tend to divide risk between Contractor and Employer. There is provision to change this responsibility in the Schedule of Rates. Clause 3.5 states that the Schedule of Rates takes precedent over Clause 3.4. If the Schedule of Rates is a published Schedule, then this aspect will need to be covered by the Site-specific Particular Clauses to the Contract.

Clause 3.6 to 3.7 (Variations)

There is provision for orders to be varied, again in writing. However, if the Contractor carries out variations that have not been given in writing, Clause 3.6.2 permits the Contract Administrator to sanction the variations. Thus, the Contractor can be given an oral instruction, if necessary, that can then be confirmed in detail when he has completed the work. This procedure could be applied to emergency work. Whichever way the Order is issued it must end up in writing. This is essential to provide the documentation for measurement and valuation.

Clause 3.8 (Cancellation of an Order)

This Clause enables the Contract Administrator to cancel an Order. The Contractor is entitled to be paid for all work carried out up to cancellation and reimbursement of any consequent direct costs incurred.

Clause 3.9 (Exclusion from the Site)

This Clause gives the Contract Administrator the power to exclude any person employed on the Site. However, he must be able to justify his actions as being reasonable in the circumstances. This power is generally exercised for safety and security reasons. The Contract Administrator would be well advised to record his reasons in writing, contemporaneously with the event, and to sign, date and keep this document.

Clause 3.10 (Non-compliance)

This is a normal condition for non-compliance with instructions. If the Contractor does not comply within seven days following receipt of a notice, the Contract Administrator can have the work done by others and deduct the cost from monies due to the Contractor.

Section 4   Payment

Clause 4.1 to 4.3 (Valuation – measurement)

It is the Schedule of Rates that is at the heart of valuing the work carried out by the Contractor. Clauses 4.1 to 4.3 describe the priority of methods of valuation of an order:

measurement and valuation in accordance with the principles of measurement used and the rates or prices in the Schedule of Rates, fair rates and prices deduced from them, or, failing that, by agreement between the parties.

As a matter of law, the test for applicability of a particular rate or price in the Schedule of Rates is: 'Is the work content of that instant rate or price executed under similar terms and conditions as the Rate contained within the Schedule of Rates?' The economic efficacy of the rate to the Contractor is not a relevant consideration.

Clause 4.4 to 4.5 (Valuation – Daywork notice and returns by Contractor)
The Contractor must give reasonable notice before he carries out work on a Daywork basis. He must provide normal Daywork returns within seven days of the end of the week in which it was carried out. The Contract Administrator, if he approves it, must sign and return the Daywork account. However, it is always up to the Contract Administrator to decide whether Daywork is the appropriate basis for valuing an order. Submission of Daywork returns does not commit the Contract Administrator to this method of valuation. He has the power to insist the Order is valued on the basis of the Schedule of Rates.

The Royal Institution of Chartered Surveyors' (RICS) Definition of the Prime Cost of Building Works of a Jobbing or Maintenance Character has been used as the basis of the calculation of the value of Daywork. However, it may be more appropriate to use the RICS Definition for Building Work. It is important to make sure that the basis on which the Daywork is to be paid is appropriate and clear.

The jobbing work definition is intended for use where the whole of the work on one contract is carried out on a Daywork basis; the Daywork labour rate varies with each specific operative, his employment conditions and the Contractor's operating costs. The Daywork labour rate used in conjunction with the RICS Definition for Building Work will provide a rate applicable to all operatives; this standard, hourly base rate is calculated and published by the Building Cost Information Service (BCIS), a company of the Royal Institution of Chartered Surveyors (RICS).

There is also the option to insert an hourly rate or ask the Contractor to quote an hourly rate at Item 9(d) of the Appendix.

The Contract Administrator would be well advised to, in conjunction with the Contractor, organise and carry out weekly, ongoing Daywork payment review meetings, the function of which would be:

1. to look back at the latest batch of submitted Daywork sheets
2. to air any queries from either side
3. to address and answer such queries whilst both sets of minds are still fresh, and
4. to record once and for all subsequent agreements.

Clause 4.6 (Overtime work)
This Clause deals with the method of payment for work carried out outside normal working hours. It defines 'overtime', 'normal working hours' and 'non-productive time' and sets out the conditions such payments are permitted.

Overtime work must be specifically directed, in writing, by the Contract Administrator and returns submitted by the Contractor in a specified format. The Contractor is then entitled to be paid the amount of the non-productive overtime rates paid by the Contractor plus the Percentage 'C' inserted in the Appendix at Item 9(f). All other overtime worked by the

Contractor and not requested in writing by the Contract Administrator is deemed to be at the Contractor's convenience and cost.

Clause 4.7 (Interruption of work – unproductive costs)
Occasionally, the Contractor may be required by the Contract Administrator to interrupt work on one Order to carry out urgent work elsewhere. In such cases, any agreed lost time or other unproductive costs are valued on a Daywork basis.

Clause 4.8 to 4.10 (Responsibility for measurement and valuation)
This Clause defines the division of responsibility for measurement and valuation. The Appendix, Item 10(a), provides for the insertion of an estimated value of an Order. Above this value, the Contract Administrator is responsible; below it, the Contractor is responsible. Where all measurement and valuation is to be undertaken by the Contract Administrator, the amount entered must be 'Nil', whereas, if Item 10(a) is deleted, all measurement and valuation is undertaken by the Contractor.

Clause 4.11 (Progress payments – Orders)
The Contractor may apply to the Contract Administrator for a progress payment, at intervals of not less than one month, providing the Contract Administrator's estimate of an Order is above the figure stated by the Employer in the Appendix to the Contract. This is subject to a minimum value of £1,500, and/or the duration of the work being 45 days or more. If the estimated value of the order, including any variations, is in excess of the figure then the Contractor is entitled to apply for a progress payment. However, it is his responsibility to make such an application. The Contract Administrator has 14 days to certify payment.

Clause 4.12 to 4.14 (Final payment – Orders)
When an Order is issued, there is an instruction to complete by a certain date, but there must also be an actual completion date. The Contractor notifies the Contract Administrator when in his opinion an Order is complete and this date stands unless the Contract Administrator dissents within 14 days.

If the Contract Administrator disagrees within the 14 days, he can either give his own completion date or he can state why he is unable to state the completion date. In practice, it is common for the Contractor and the Contract Administrator to sign off the Order jointly. The payment process is dependent on the actual completion date and, therefore, it is essential that each Order has one.

The measurement and valuation of Orders is either by the Contract Administrator (Clause 4.13 pursuant to 4.9), by the Contractor (Clause 4.14 pursuant to 4.8 or 4.10), or can be split so that the Contractor measures Orders below a certain value.

When measurement is by the Contract Administrator, he must certify the value within 56 days of completion of the Order. This may seem a long period, but it is considered far better for the Contract to include a period attainable in most circumstances.

After 56 days, if the Contract Administrator is responsible for the measurement and he has not done it, the Contractor can give him notice and after 28 days he can carry out the measurement himself. This measurement is given to the Contract Administrator and if within 28 days he does not object to it, it becomes the value of the Order.

When measurement is by the Contractor, he also has 56 days from completion of an Order to submit the account. The Contract Administrator then has 28 days in which to check it and certify it. If the Contract Administrator finds errors, he has a right to deduct the cost of finding those errors from the Contractor's payment. This condition is rarely applied in practice, but if there is consistent over-measurement by the Contractor, then the provision to impose this clause means the cost of re-measuring the Work Order value can be recovered from the Contractor. This gives the Contract Administrator a mechanism to deter consistent over-measurement by the Contractor.

If the Contractor has not provided measurement within 56 days, there is provision for the Contract Administrator to give notice that after 28 days he can do the measurement himself and deduct the cost from the Contractor. The Contractor does not have the right to charge the Employer the cost of measurement if he fails to do it, but the Contract Administrator has the right to charge the Contractor. This may seem inequitable, but the Contractor will at least have the right to value the Order and, even without payment for measurement, will receive payment for the Order itself.

Clause 4.15 (Payment of certificates)
Certificates must be paid by the Employer within 14 days of the issue of a certificate by the Contract Administrator.

There is a fairly long period between work being carried out and payment. Clauses 4.13 and 4.14 specify 56 days for measuring and valuing the Order, 28 days to check it and certify it, and then 14 days to pay. The times given in the Contract are maximum times, however, and most Employers would be expected to make payments well within those times.

Clause 4.16 (Right of suspension of obligations by Contractor)
If the Employer fails to pay the Contractor in full by the final date for payment, the Contractor may give the Employer written notice of his intention to suspend his obligations under the Contract. If the failure continues for seven days after the notice, he has the right to carry out the suspension.

Section 5 Statutory obligations

Clause 5.1 to 5.4 (Statutory obligations, notices, fees and charges)
This Condition describes the Contractor's obligations to give all statutory notices and pay all fees and charges in connection with all legislation and, in particular, the Construction (Design and Management) Regulations 1994 (SI 1994 No. 3140). Fees and charges are reimbursed to the Contractor, but there is no provision for reimbursing any other costs or overheads in connection with compliance. The Contractor is deemed to have provided for this in his tender.

Section 6 Injury, damage and insurance

Clause 6.1 to 6.4 (Injury to persons and property and indemnity to Employer)
The Contractor indemnifies the Employer against any claims for personal injury to, or death of, third parties other than where it is the Employer's fault, and damage to third-party property where it is caused by the Contractor's negligence or default. However, the indemnity for damage to third-party property does not apply to Orders.

The Contractor must also maintain insurance for injury to persons and damage to property.

Clause 6.5 to 6.7 (Insurance of existing properties)
There is a requirement for the Employer to insure all the existing structures in the Contract Area. The Employer must insure all these buildings for specified perils, notify his insurer that work may be carried out in them and have the Contractor as a joint insured or the rights of subrogation against the Contractor must be waived.

Clause 6.8 to 6.13 (All Risks Insurance)
To cover the work comprised in the Orders, the Contractor must maintain a Joint Names Policy for All Risks Insurance, for the full reinstatement value of each and every Order up to its completion date. Joint names' policies obviate later disagreements between the Employer and the Contractor as to liability.

Section 7 Determination

Clause 7.1 to 7.2 (Determination by Employer)
This condition gives the Employer the power to determine the employment of the Contractor under the Contract if he fails to comply with the CDM Regulations or with his obligations under the Contract. He issues a notice to the Contractor and if the Contractor does not comply within 14 days, he may determine.

Clause 7.2 covers automatic determination through bankruptcy or insolvency.

Clause 7.3 (Corruption – determination by Employer)
The Employer has the power to determine the Contract if he is reasonably satisfied that corruption has taken place, even if it was carried out by one of the Contractor's employees without the Contractor's knowledge.

Clause 7.4 to 7.5 (Determination by Contractor)
If the Employer fails to comply with his obligations under the Contract, does not pay by the final date for payment, or fails to comply with the CDM Regulations, the Contractor may determine his employment under the Contract.

Clause 7.5 reflects 7.2 above and covers automatic determination through bankruptcy or insolvency of the Employer.

Clause 7.6 to 7.7 (Provisions consequent on determination under clause 7.1, 7.2, 7.3, 7.4 or 7.5)
Clause 7.6 provides for the valuation of work not valued and certified before determination to be certified in accordance with Section 4. Clause 7.7 permits the Employer to withhold any direct loss and/or expense caused to the Employer by the determination.

Section 8 Break provisions – Employer or Contractor

Clause 8.1 provides a 'break' clause enabling either party to terminate the Contract by giving 13 weeks' notice, provided this is not within 6 months of the commencement of the Contract Period. The Contractor is under no obligation to carry out any Orders issued after the notice has been given that cannot reasonably be completed by the expiration of the notice. However, he must complete all Orders received prior to the notice, even if they cannot be completed by the expiration of the notice.

Section 9 Settlement of disputes

9A Adjudication
Clause 9A.1 (Application of clause 9A): This Clause applies if, in accordance with Article 6, any dispute arises under the Contract and either party refers it to adjudication.

Clause 9A.2 (Identity of Adjudicator): The Adjudicator to decide the dispute is either an individual agreed by the parties, or an individual nominated by the person named in the Appendix.

Clause 9A.3 (Death of Adjudicator – inability to adjudicate): This Clause deals with a situation where the nominated Adjudicator is unable to adjudicate.

Clause 9A.4 (Dispute or difference – notice of intention to refer to adjudication-referral): This Clause sets out the procedure for referring the dispute to the Adjudicator ('the referral'). The Party requiring the referral must include particulars of the dispute, a summary of the contentions, a statement of the relief or remedy sought and any material to be considered.

Clause 9A.5 (Conduct of the adjudication): Once the Adjudicator has received the referral, he confirms receipt to the Parties. Clause 9A.5 then goes on to describe the conduct of the adjudication.

Clause 9A.6 (Adjudicator's fee and reasonable expenses – payment): In his decision, the Adjudicator states how payment of his fee and reasonable expenses are to be apportioned between the parties.

Clause 9A.7 (Effect of Adjudicator's decision): The decision of the Adjudicator is binding on the Parties until the dispute is finally determined by arbitration or by legal proceedings. Alternatively, the Parties may agree to accept the decision of the Adjudicator.

The arbitration or legal proceedings are not an appeal against the decision of the Adjudicator, but are a consideration of the dispute as if no decision had been made by the Adjudicator.

Clause 9A.8 (Immunity): This Clause grants immunity to the Adjudicator for anything he does, or omits to do, unless he has acted in bad faith. This immunity also extends to the Adjudicator's employees or agents.

9B Arbitration
Clause 9B.1 (Notice of reference to arbitration): This Clause applies if, in accordance with Article 7A, any dispute arises under the Contract and either party serves a notice of Arbitration.

Clause 9B.2 (Powers of Arbitrator): The Arbitrator has greater powers than the Adjudicator, being able to rectify the Contract so that it accurately reflects the true agreement made by the parties.

Clause 9B.3 (Effect of the award): The award by the Arbitrator is final and binding on the Parties.

Clause 9B.4 (Appeal – question of law): This Clause allows either Party to apply to the courts on a question of law arising from the reference to arbitration or the award made by the Arbitrator.

Clause 9B.5 (Arbitration Act 1996): The provisions of the Act apply.

Clause 9B.6 (Conduct of arbitration): This Clause describes the model arbitration rules to be followed in conducting the arbitration.

GC/Works/7 (1999) Measured Term Contract

GC/Works/7 (1999) is a new edition of the standard Government form of contract for a Measured Term Contract replacing the General Conditions of Contract for Measured Term Contract Form C1501. Although it is aimed at government departments, it is suitable for most public or private sector organisations. If used in the private sector, the most significant adjustment would be the provision of conditions covering insurance. The following is a brief summary of the Conditions, but should not be regarded as a substitute for reading the Contract itself.

Condition 1 (Definitions, etc.)

Paragraph (1) gives the definitions of terms to be used throughout the Contract (not just the Conditions). Most of the definitions are self-explanatory. Some, such as 'the Contract Area' and the 'Contract Period' are cross-referenced with the Abstract of Particulars. It is essential that such information be inserted if the Contractor is to be expected to submit a reasonable tender. Defined terms are denoted, following accepted legal terminology, by the use of initial capital letters.

Condition 2 (Fair dealing)

This Condition reflects the current trend towards 'partnering' in that it goes beyond a strict adherence to the letter of the Contract and imposes an obligation to deal fairly, in good faith and with mutual co-operation. Measured Term Contracts require a high degree of flexibility from both parties (the much quoted 'swings and roundabouts' approach) and this Condition attempts to formalise this. This Condition is, of necessity, written in innocuous language, but has been held as a matter of how to influence the interpretation of all other Conditions of Contract.

Condition 3 (Contract documents)

This Condition deals with the priority of Contract documents. In general, any Supplementary Conditions and Annexes prescribed by the Abstract of Particulars have priority over the General Conditions, but the General Conditions will prevail over all other documents forming part of the Contract.

Condition 4 (Delegations and representations)

There may be numerous individuals with an interest in repairs or maintenance ranging from irate occupiers to harassed ordering clerks. This Condition ensures that it is only the Project Manager (PM) who is authorised to communicate decisions on behalf of the Employer to the Contractor. If the Employer wishes to reserve any contractual powers to himself, he must do so under Condition 1(1), Project Manager, in the Abstract of Particulars (Model Form 1). The Employer has to name the person who will exercise these reserved powers.

Condition 5 (Supervision and attendance)

Good communication and supervision are vital to the efficient running of a term contract. This Condition requires a competent and accessible agent to be appointed by the Contractor to receive Orders and Instructions from the Project Manager. The Condition does not specify that the agent is to be site-based, but on a Contract of a significant size this may be desirable.

Condition 6 (Contractor's employees)

This Condition gives the Project Manager extensive powers to require the Contractor to remove any person whose continued employment is, in his opinion, 'undesirable'. There is no specific requirement for the Project Manager to give reasons, but Condition 2 (Fair dealing) implies a reasonable application of these powers and a need to discuss the reasons openly with the Contractor's agent and record all available evidence. The Project Manager would be well advised to record such evidence contemporaneously with the exercise of his powers, and to sign, date and keep this evidence.

Condition 7 (Scope of the Contract)

This Condition gives the Contractor's responsibilities under the Contract and provides a 'break' clause enabling either party to terminate the Contract by giving three months' notice.

Condition 8 (Orders and Instructions)

This Condition describes the process of placing an Order with, or giving an instruction to, the Contractor. Instructions and Orders should normally be in writing, but emergency Instructions and Orders may be given orally and confirmed in writing within seven days.

Paragraph (2) is specific about the information the Order must contain, including:

- the date of commencement and date for completion of the Order
- whether the Employer will provide any materials
- whether the work will be measured jointly by the Contractor and the Quantity Surveyor (QS), and
- the Project Manager's estimate of the value of the Order.

The use of a standard format for Orders would be a distinct advantage (for example, see Model Form 6). Inadequate ordering information is a frequent cause of problems on term contracts.

If the Contractor is unable or unwilling to promptly carry out an Order or Instruction, paragraph (4) enables the Project Manager to arrange for a third party to carry out the work and recover the cost and any reasonable expenses from the Contractor.

Condition 9 (Defects in Maintenance Period)

Condition 9 requires the Contractor to rectify any defects appearing during the Maintenance Period that the Employer considers the Contractor's fault. The onus is on the Contractor to show that any defects are not down to him, in which case the Employer reimburses the Contractor for any costs incurred. Paragraph (3) enables the Employer, if the Contractor defaults, to execute the remedial works at the Contractor's expense. Condition 24 (Quality) deals with defects that become apparent in the course of the work.

Condition 10 (Statutory notices and CDM Regulations)

This Condition describes the Contractor's obligations to give all statutory notices and pay all fees and charges in connection with all legislation and, in particular, the Construction (Design and Management) Regulations 1994. Fees and charges are reimbursed to the Contractor (paragraph (3)), but there is no provision for reimbursing any other costs or overheads in connection with compliance. The Contractor is deemed to have provided for this in his tender.

Condition 11 (Intellectual property rights)

This Condition requires the Contractor to pay any royalty, licence fee or other expense in connection with the work, and reimburse the Employer if the use of any intellectual property gives rise to any claim or proceedings against the Employer. The Employer reimburses the Contractor if the cost was necessary in order to comply with an instruction, provided it was not reasonably contemplated under the Contract.

Condition 12 (Protection of Works)

This Condition requires the Contractor to take reasonable measures to maintain security and safety, to comply with statutory regulations for storage and to keep the site tidy.

Condition 13 (Nuisance)

This Condition requires the Contractor to take all reasonable precautions to prevent nuisance or inconvenience to tenants, occupiers or the general public. These terms make the creation of a nuisance by the Contractor a breach of Contract, besides being a possible breach of the general or criminal law.

Condition 14 (Loss or damage)

This Condition defines the Contractor's liability to make good any loss or damage arising out of, or in any way connected with the execution of the Works, including loss or damage to third-party property, personal injuries, sickness, death, loss of profit or loss of use. As Government Departments do not, as a matter of policy, carry insurance, this is a very important Condition for the Employer.

Condition 15 (Occupier's rules and regulations)

This Condition places an obligation on the Contractor to ascertain the rules and regulations that apply in respect of the Site or part of the Site at the date of the Contract. Government Contracts are used frequently on high-security establishments such as military bases. The Contractor must assess any known additional costs, such as delays to access for security checks, and include them in his tender. Post-contract changes to Occupier's rules and regulations should be notified by the Project Manager under Condition 8.

Condition 16 (Discrimination)

This Condition places an obligation on the Contractor to ensure that he, his employees, agents and sub-contractors do not unlawfully discriminate within the meaning and scope of legislation. It is common practice for Employers to obtain, at tender stage, a declaration of policy on discrimination from all tenderers.

Condition 17 (Corruption)

This Condition prohibits any corrupt actions by the Contractor, specifically bribery or the payment of commission without disclosure. The Employer has the power to determine the Contract if he is reasonably satisfied that corruption has taken place, even if it was carried out by one of the Contractor's employees without the Contractor's knowledge, or there has been no criminal prosecution.

Condition 18 (Records)

This Condition requires the Contractor to maintain such records as may be reasonably necessary for the Project Manager or Quantity Surveyor to administer the Contract. It is in the Contractor's interest to keep good records in order to justify his claims for payment. Poor record-keeping often reflects poor management, consequent poor workmanship and sometimes indicates that the Contractor has something to hide.

A contractually knowledgeable Employer would not passively rely on the keeping of records solely by the Contractor. Putting it at its mildest, it is not within the Contractor's commercial interests to record his own deficiencies, failings and inefficiencies. If the Employer wishes, at a later date, to cut out nugatory costs presented by the Contractor, he must have precise, and better still, agreed records of all such Contractor defaults. Without prejudice to the above, the Employer should, during the Contract Period, check that the Contractor is keeping accurate and detailed records.

The optimum situation with regard to record-keeping is contemporaneous agreed, signed and dated joint records, kept by both the Employer and the Contractor. This cuts out any later acrimonious arguments concerning what actually happened on any given date and diminishes adversarial conduct.

Condition 19 (Site admittance)

This Condition requires the Contractor to take all reasonable steps to prevent unauthorised persons being admitted to the Site and to provide a list of names, addresses and such other particulars as the Project Manager may reasonably require. It is common practice for a full security clearance to be required for high-security establishments. It should be noted that paragraph (2) applies to 'persons concerned with the Works or any part of them' and is not merely confined to those present on Site.

Condition 20 (Passes)

This Condition gives instructions for the issue and return of passes where they are required for admission to the Site. Passes are issued by the Project Manager. Typical delays in issuing passes should be spelt out in the Contract Documents.

Condition 21 (Photographs)

This Condition prohibits the taking of photographs of the Site or Works by the Contractor. Again, this reflects the secure nature of Government premises. Before the Project Manager issues a blanket prohibition on the taking of photographs, he should think about the Contractor's duties under Condition 18.

Condition 22 (Official secrets and confidentiality)

This Condition places an obligation to take all reasonable steps to ensure that all persons employed by the Contractor or his sub-contractors in connection with the Contract are aware of the Official Secrets Act 1989 and any other appropriate Acts. By use of the widely drafted words 'in connection with the Contract', this would include any off-site personnel.

Condition 23 (Vesting)

This Condition aims to ensure that all materials to be used in the Works (described as 'Things for incorporation', e.g., bricks, doors, windows) become the property of the Employer as a safeguard against the Contractor failing to complete the Works due to insolvency or the like. This is qualified, of course, by the requirement that the Contractor must be able to transfer ownership. In other words, he must own the materials in the first place before ownership can be transferred to the Employer. This Condition also applies to Contractor-owned plant.

To protect this right, paragraph (4) prohibits the removal of 'Things' without the permission of the Project Manager.

Condition 24 (Quality)

This Condition requires the Contractor to execute the Works with diligence, with all reasonable skill and care and in a workmanlike manner. It gives the Project Manager powers to reject unsatisfactory work or materials and requires the Contractor to draw the Project Manager's attention to any 'Things' that, in his opinion, should not be incorporated in the Works. The Project Manager's right to test 'Things' are laid out in paragraph (4).

Condition 25 (Excavations)

This Condition covers the ownership of materials and objects uncovered during excavations, demolition or dismantling. They remain or become the property of the Employer. This can include materials for re-use, fossils or antiquities. Paragraphs (3) and (4) also deal with the safe handling of fossils and antiquities.

Condition 26 (Measurement)

Once an Order has been carried out, the work will need to be measured. This Condition gives the rules for arranging this. The Contractor must give the Quantity Surveyor 14 days' notice of the day and time he intends to measure. The Quantity Surveyor then has the discretion as to whether he attends unless the Project Manager instructs that the measurement is carried out jointly by the Quantity Surveyor and the Contractor.

Paragraph (2) requires the work to be valued in accordance with Condition 27. Paragraph (3) empowers the Project Manager to instruct the Contractor to correct any errors or unfair prices or rates. If the Contractor fails to keep up with the measurement of Orders, paragraph (5) allows the Project Manager to instruct the Quantity Surveyor to measure and value independently, although paragraph (6) gives the Contractor the opportunity to object by issuing a notice to the Project Manager within seven days.

Condition 27 (Valuation)

Following on from Condition 26 (Measurement), this Condition gives the priority for valuation of the work. The expected method is by using the net rates and prices in the Schedule of Rates, adjusted by the Updating Percentage Adjustment (although, in practice, this is sometimes deleted from the Contract) and the Contractor's Percentage Adjustment. There are bound to be occasions where there are no appropriate rates in the Schedule of Rates in which case, in order of priority, rates or prices deduced from the Schedule (adjusted as above), 'fair' rates and prices or Daywork, should be used.

The method of dealing with the Contractor's profit in respect of prime cost items or where materials are supplied which are not included in the Schedule of Rates is also described.

As a matter of law, the test for applicability of a particular rate or price in the Schedule of Rates is: 'Is the work content of that instant rate or price executed under similar terms and conditions as the Rate contained within the Schedule of Rates?' The economic efficacy of the rate to the Contractor is not a relevant consideration.

The Contract Administrator, in conjunction with the Contractor, would be well advised to organise and carry out weekly, ongoing daywork payment review meetings, whose function would be to:

1. look back at the latest batch of submitted Daywork sheets
2. to air any queries from either side
3. to address and answer such queries whilst both sets of minds are still fresh, and
4. to record once and for all subsequent agreements.

Condition 28 (VAT)

This Condition deals with the issue of Value Added Tax, describes the Contractor's obligations with regard to VAT and gives the Employer powers to recover any additional VAT he may incur through the Contractor's failure to fulfil those obligations.

Condition 29 (Advances on account)

Cash flow is extremely important to the Contractor to enable him to finance the Contract. This Condition gives the circumstances under which, by exception, the Contractor is entitled to be paid an advance on account for work on an Order in progress. To reduce the administrative burden, this is limited to Orders whose values are in excess of a value stated in the Abstract of Particulars to the Contract and/or the duration of which is 45 days or more.

It should be noted that there is no provision for retention to be held.

Condition 30 (Final Payment)

On completion of an Order, the Contractor is entitled to be paid the full value as certified by the Project Manager. Any advances that exceed the final payment certified must be adjusted and the excess repaid to the Employer.

Condition 31 (Certifying payment)

Subject to certain Conditions, the Project Manager certifies payment to the Employer with a copy to the Contractor. Certification must take place within 14 days of the Contractor's valuation, his application for an advance on account, a joint measurement carried out in accordance with Condition 26(3) or a measurement and valuation carried out by the Quantity Surveyor in accordance with Condition 26(5).

Only when the certificate has been issued can the Contractor submit his invoice. The Employer must then pay within 30 days of receipt.

Paragraph (5) is a warning that no certificate is final and conclusive evidence. Any certificate may be modified or corrected.

Condition 32 (Payment notification and withholding payment)
This Condition gives the course of action if either party fails to carry out his obligations under the Contract with regard to payment. Not more than five days' notice is given after the date on which payment was due. If either party intends to withhold payment of a sum due, 'effective' notice must be given not later than seven days before the final payment of the sum. To be 'effective', the notice must specify the amount it is proposed to withhold and the grounds for withholding each separate payment.

Condition 33 (Prolongation and disruption)
The Contractor may claim for 'properly and directly' incurred expense, but not indirect expense, under this condition. Paragraph (1) gives the grounds for a claim, e.g., as a result of other works being executed on the Site (see Condition 43 (Other works)), delay and advice given in accordance with the CDM Regulations. Paragraph (2) expands on delay other than delay in being given access to the Site.

Paragraph (3) is expressed strongly as a Condition precedent; in other words, non-compliance on the Contractor's part will disbar him from receiving any expense.

Paragraph (3) also requires the Contractor to give notice to the Project Manager that he expects to be entitled to an increase in the final sum of an Order as a result of regular progress being, or likely to be, disrupted or prolonged. This notice provision is designed to alert the Project Manager that the Contractor is incurring expense, so that the Project Manager can take possible mitigating action as soon as possible. He must then provide, within 56 days, full details of his expense and evidence that the expense was incurred directly as a result of the events in paragraph (1).

Condition 34 (Recovery of sums)
This Condition gives the Employer additional rights to deduct any sums recoverable from sums due under the Contract or any other contract between the Contractor and the Employer.

Condition 35 (Dayworks)
This Condition requires the Contractor to give reasonable notice of the start of any work to be executed by Daywork and to provide details in the required form within seven days of the end of each week.

The Project Manager is not obliged to accept the Daywork vouchers if, in his opinion, they appear unreasonable. To make this judgement, the Project Manager must keep good records. He may instruct the Quantity Surveyor to reduce the time and quantities or, if the Project Manager considers that the work could be valued using contract rates, to measure and value in the normal way.

Condition 36 (Suspension for non-payment)
This Condition defines the rights of both parties to suspend performance of their obligations under the Contract should a sum due not be paid and notice under Condition 32 (Payment notification and withholding payment) has not been given. At least seven days' notice of intention to suspend performance must be given and the right to suspend performance ceases when payment is made.

Condition 37 (Determination by Employer)

This Condition gives the Employer the power to determine the Contract under certain defined grounds. The grounds are:

- failure to carry out the Works in accordance with the Contract
- failure to comply with an Order or Instruction within a reasonable time
- the Contractor becoming insolvent
- failure to satisfy the Employer with regard to Condition 19 (Site admittance)
- corruption as defined in Condition 17 (Corruption), or
- any breach of conditions in the invitation to tender.

The remainder of the Condition defines insolvency for a company (paragraph (2)), a partnership (paragraph (3)) and an individual (paragraph (5)).

Condition 38 (Consequences of determination by Employer)

Following on from Condition 37 (Determination by Employer), this condition deals with the calculation of the sums due to the Contractor. All sums due are suspended until the work has been completed by another Contractor and the amount due to the original Contractor has been calculated after recovery of the Employer's costs incurred.

Condition 39 (Adjudication)

A history of litigation in the construction industry has led to the increased use of adjudicators to settle disputes at an early stage. This Condition gives either party the opportunity to refer disputes to a person named as adjudicator in the Abstract of Particulars.

The Condition gives detailed information on the procedure of adjudication and the powers of the adjudicator.

Condition 40 (Choice of law)

This Condition gives a choice of law (English law in England or Wales, Scottish law in Scotland or Northern Ireland law in Northern Ireland) according to the location of the Works.

Condition 41 (Assignment)

This Condition prohibits the Contractor from assigning or transferring the Contract without the Project Manager's consent, but the Employer may do so without the Contractor's consent.

Condition 42 (Sub-letting)

The Contractor cannot sub-let any part of the Contract without the consent in writing of the Project Manager, but it can be inferred that this consent cannot be unreasonably withheld (Condition 2). In practice, the wide range of work that could occur on a maintenance and repair contract is bound to involve the use of specialist sub-contractors. This is usually enabled pre-contract by the supply by the Contractor of his list of proposed sub-contractors, which the Project Manager approves in writing. Paragraphs (2) and (4) make it clear that the Contractor is responsible for any sub-contractor employed by him.

Condition 43 (Other works)

This Condition recognises that the Employer may wish to execute other works on the Site. The Contractor must 'give facilities' for these works, but Condition 33 (Prolongation and disruption) acknowledges that the Contractor may 'properly and directly' incur expense as a result and consequently be entitled to an increase in the final sum due to him.

Paragraph (2) makes the Employer responsible for damage to other works for the purposes of Condition 14 (Loss or damage) except where caused by the negligence, omission or default of the Contractor's workpeople, agents or sub-contractors.

Table 9.1   Comparison of Contract Forms

Subject	JCT Standard Form (1998)	GC/Works/7 (1999)
Employer's representative	Contract Administrator – Articles of Agreement	Project Manager – Abstract of Particulars (Model Form 1)
	Quantity Surveyor – not applicable	Quantity Surveyor – Condition 1
Contract Area	Articles of Agreement and Item 1(a) of the Appendix	Abstract of Particulars (Model Form 1)
Type of Work	Item 1(b) of the Appendix	No separate provision
Order values	Item 2 of the Appendix	Abstract of Particulars (Model Form 1)
Approximate anticipated value of work	Clause 1.13 and Item 3 of the Appendix	No separate provision
Contract Period	Item 4 of the Appendix	Abstract of Particulars (Model Form 1)
Priority coding for Orders	Item 4 of the Appendix	No separate provision
Schedule of Rates	Item 6 of the Appendix	Item (e) of the Tender Form (Model Form 3)
Fluctuations	Item 7 of the Appendix (Item 8 for Fixed Price)	Item (f) of the Tender Form (Model Form 3)
Dayworks Provision	Item 9 of the Appendix	Item (g) of the Tender Form (Model Form 3) and Contract Daywork Schedule (Model Form 4)
Defects	Clause 2.4 – 6 months	Abstract of Particulars (Model Form 1)
Responsibility for Measurement and Valuation	Item 10(a) of the Appendix	Condition 26
Progress Payments	Item 10(b) of the Appendix	Abstract of Particulars (Model Form 1)
Insurance	Item 13 of the Appendix	No separate provision
Break Provision	Clause 8.1, Item 14 of the Appendix – default 13 weeks	Condition 7.2 – 3 months
Settlement of Disputes	Item 15 and 16 of the Appendix	Abstract of Particulars (Model Form 1)

10 Computerised Measured Term Contract administration

Introduction

A Schedule of Rates computer system can enable Employers and Contractors working on Measured Term Contracts to process Works Orders and Estimates with the minimum of input. Compared to manual methods, computerised administration offers substantial savings in time and accuracy, as well as an improvement in presentation.

One of the main benefits is to dramatically reduce the time spent by surveyors writing out and calculating the value of a Works Order or Estimate, including squaring dimensions, extending them by a rate and applying Contract percentages. By retaining the Order details, an Order Ledger is created and various reports on the Contract can be generated automatically.

Many Contracts can be held and new ones are added with the minimum of effort. Contract information held can include Contract number, description, year of Schedule of Rates in use, Contractors percentage(s), etc. Besides any published Schedule of Rates, Addendum and Bespoke Schedules can also be added, including importing Schedule details directly from Microsoft Excel or ASCII format.

Works Orders and Estimates

New Works Orders and Estimates are easily created. Details such as Order description and date of issue are entered and the relevant Item reference numbers can either be selected directly from the Schedule of Rates by displaying the appropriate section on screen or the items can be typed in. Dimensions or total quantity required can then be entered. When entry of an Order is complete, it can either be printed out or displayed on screen.

When a printout of an Order is produced, each item number and description is listed, the dimensions are squared to produce the quantity and this is extended by the rate. The total price appears and is adjusted by the Contractor's percentage(s) and, if applicable, the Monthly Update percentage(s).

Besides Schedule of Rates items, non-Schedule items such as agreed rates are easily added to an Order. There are various options that enable pro-rata or agreed rates, net rates, Dayworks, supply only, invoices, sub-contractors, materials, hired plant, PC Sums and free-typed descriptions to be entered. Each requires a description entry and, except for free-type, a unit of measure and rate.

At any time, Orders are easily amended and mistakes rectified. Missing Item numbers or dimensions can be added and incorrect dimensions altered.

Additionally, notes can be stored against an Order. These are intended for reference and can be updated at any time. The notes could be a record of activity on an Order, such as dates of visit to site and observations on the work. The intention is to replace manual records for each Order.

The Order Ledger

An Order Ledger is maintained for each Contract. The ledger holds details of all Orders entered including Order description, date of issue, measured value, various other dates, property register details, etc. This enables many reports to be produced for a Contract, including a list of all Orders, Outstanding Orders and a Property Register, which could include activity for a selected building or room within. There is also the ability to export the Order Ledger information directly to Microsoft Excel.

Estimating

Estimating, too, is made easier as the Schedule of Rates can be displayed on-screen, instead of referring to the printed Schedule, and the Item numbers required can be quickly located. This technique can also be used when entering Orders. It is achieved by displaying different levels of description to enable rapid identification of the code required. For example, with the PSA Schedule of Rates for Building Works, firstly SMM7 work groups are displayed, e.g. D: Groundwork, E: In-Situ Concrete. A work group is selected and the associated work sections then appear, e.g. E10: In-Situ concrete, E20: Formwork for in-situ concrete. When the work section is chosen, all the constituent work heads are displayed. Selection of a work head enables the relevant Item numbers and descriptions to be displayed, such that an item can be selected for the estimate. Further Item numbers can be selected by continuing to search through work groups, work sections and work heads. This is a very convenient and quick way of locating codes, i.e., rather than searching through the printed Schedule. Other methods of searching include displaying the printed Index of the Schedule of Rates and selecting the appropriate Item numbers, or a word or phrase can be searched for and all codes containing that word or phrase will be displayed.

Composite items

Composite items are an option with computerised systems. They offer great savings provided that they are used in the correct situation and agreed with the Employer. Composite items can be created from a group of items to form a complete component installation, such as fitting a standard window or door. When the composite is created, each item number must have standard dimensions entered. The entry of a composite Item into an Order or Estimate saves the selection of each constituent Item number and its dimensions. The dimensions on the composite may not be exactly correct but, as long as the composite is reasonably similar to what is required, the amount of time saved outweighs any loss of accuracy.

If required, each Order can be referenced by up to three different Order Numbers. The three could be the official Employer's Order Number, the Contractor's internal Order Number and an automatically generated sequential Order Number. The Order can be immediately referenced using any of these numbers.

The systems are generally written in Visual Basic, the popular and powerful development tool from Microsoft. Each Schedule of Rates is stored in a Microsoft Access database. This ensures rapid response and ease of use, even when large amounts of data are involved. If the User's computer system has the Access database package loaded, it can be used to interrogate the data generated by the computerised Schedule of Rates and produce reports with reference to the Order Ledger. However, Access does not have to be loaded on the computer. The software will operate across the various versions of Windows available including Networking.

Why use computerised Schedules of Rates?

The volume of calculations required when using a Schedule of Rates on a Measured Term Contract makes it an obvious choice for the application of computerisation. Whether you just purchase a Schedule of Rates as data and use it with a standard database or spreadsheet, such as Microsoft Access or Excel, or you purchase a system that has been specifically written to operate with Schedules of Rates, the benefits should be considerable. It is recommended that you use a computer package specifically written for Schedules of Rates as it is more relevant to the application. Computer packages have been designed for this use, over Access or Excel, which can carry out a multitude of applications, but in this case, are not specialised enough. There are, however, a number of organisations using Access and Excel to process Schedules of Rates to their satisfaction and, of course, they are a much cheaper option.

Choice of system

There are a number of systems available that specifically operate with Schedules of Rates, some vastly different, others very similar. When looking at systems, decide upon what exactly is required and look at those systems that cover your requirement. Remember, the aim of a computer package in a particular market is to satisfy all of the people all of the time, whilst being simple to use and producing exactly what is required.

There is a wide variety of systems available. For example, there are systems aimed at the Employer, local authority or quantity surveyor that deal with the generation of Estimates, Works Orders and the production of Work tickets, keeping track of large numbers of Orders and the control of budgets. Other systems are aimed more at Contractors and deal with the processing of Works Orders and Estimates, including having the ability to inspect the Schedule of Rates on-line and produce a Works Order by selecting a Schedule item by the click of the mouse, followed by entering the dimensions. When the Works Order is complete, the relevant percentages can be automatically applied to give a total value of that Order. An Order Ledger can be automatically created from the Works Orders that are entered and many reports showing Orders outstanding or processed within a certain period can be produced.

Contractors are usually the ones with the Schedule of Rates on-line on their computer, whilst the Employer's interest is in receiving the details of completed Works Orders from the various Contractors and to be able to analyse the results, as well as approving orders for payment. Both Employers and Contractors using the system vary in their needs: there are Employers and Contractors operating one or more small contracts, Employers with a large Contract and many Contractors, and Contractors operating many Contracts using an assortment of Schedules of Rates.

Specifying computer systems

Increasingly, Employers, local authorities, quantity surveyors, etc., are putting Contracts out to tender stating that Contractors must use a specified computer system. It is also becoming increasingly common that Contractors are being required to submit their completed Orders electronically, either by email or floppy disk, instead of as hard copy.

As new Schedules are released, they are made available and can be added to a system as the need arises. Schedules can be combined; a Contract may consist of Building, Decoration, Mechanical and Electrical Schedules of Rates, for example. Systems can be modified easily by Employers and Contractors using various Schedules of Rates on different Contracts.

Systems will operate on standard PCs, notebooks or laptops, or a Network, where users can share information. With a Network, all users are accessing the same information. When a Works Order is entered, all users can interrogate it.

Choice of operator

A computer package should be a tool to enable the user to carry out the tasks easily and efficiently. It should fit in to the established pattern of work and not involve the user in learning a completely new way of operating.

It has been found that different levels of employee operate systems according to the requirements of an office. They range from quantity surveyors through technicians to secretaries. If the Order information being entered into the computer has already been coded, rather than involving direct selection of codes on screen, then more junior members of staff can be involved. This frees the surveyor from this laborious task.

In general terms, if, before computerisation, the quantity surveyor produced hand-written dimensions which are passed to the Employer, then he may operate the computer system. However, if, when operating manually, his dimension sheets are passed to a technician or secretary, then they will be more likely to operate the computer. If using on-screen selection of codes from the Schedule of Rates, then the surveyor will be the operator.

Input to the computer can either be from dimension slips containing brief descriptions of the work carried out and the full description located on screen, or the dimension slips can be passed to the more junior member of staff for input. Either full dimensions or total quantities can be entered.

The secret of a successful system is that it fits into an office with the minimum of disruption.

Advantages and disadvantages of computerisation

The advantages and disadvantages of choosing to computerise the administration of a Measured Term Contract can be summarised as follows:

Advantages:

- Time savings, including more senior staff being able to delegate work.
- Accuracy, including knowing that the squaring of dimensions is correct, as is the item rate and any updating percentages, etc., to be applied.
- Increased job satisfaction from reductions in manual calculations.
- Creation of an Order Ledger from Works Orders entered enables many different Order reports to be generated, e.g., Overdue Orders, value of Completed Orders within any period, etc.
- Electronic transfer of information, either Employer-to-Contractor or vice versa, or from one office to another. Measures can be readily available to all as soon as they have been entered.
- Ease of altering information and regenerating the Works Order or an Order Ledger report.
- Speed of locating Schedule of Rates descriptions by clicking through headings, sub-headings, etc. or by searching for a phrase.

- Auditors and Employers will have no problem understanding the printed documents, as compared to written ones, and readily accept their accuracy.
- Greater standardisation of procedures within an office. Without standardisation, there is a danger that Monthly Update percentages and Contract percentages will be applied in different ways by different staff. The standard system operates the same every time.
- Bespoke reports can be written by accessing the information in the standard database within the system.

Disadvantages:
- Initial cost of purchasing a computer system. If a system is purchased, then the costs must be recouped from the profit within the Contract(s).
- Time spent in training staff to use the system and becoming proficient with its operation.
- Insufficient work on the Measured Term Contract to justify computerisation.
- Insufficient Contract Period remaining to justify computerisation. The system supplier may need to be consulted to see if a license for a shorter period at a reduced price can be obtained, or if software can be rented.
- Location may be a factor. Measurements can be manually written up at Site, in the office or at home.

Conclusion

A computerised Schedule of Rates computer system can enable Employers and Contractors working on Measured Term Contracts to process Works Orders and Estimates with the minimum of input. It offers substantial savings in time, greater accuracy and improved presentation over manual methods. The time spent by surveyors writing out and calculating the value of a Works Order or Estimate is dramatically reduced and by retaining the Order details, an Order Ledger is created and various reports on the Contract can be generated automatically. Many Contracts can be held and new ones added with the minimum of effort.

Introduction

This chapter draws on experience and procedures in large organisations with contracts spread over the whole of the UK. It covers matters of principle and gives just an outline of how the problem should be tackled.

Some measure of auditing is recommended to ensure probity, contract compliance, quality of work and value for money.

Background

Government departments, local authorities and other public bodies are publicly accountable for all money that they spend. Other organisations are similarly accountable; for example, private companies are usually accountable to shareholders. Public bodies and private companies alike need to be able to show that money has been spent correctly and that accounts stand up to scrutiny. They must also demonstrate, where possible, that value for money has been achieved.

Audit procedures should reflect the constraint (or discipline) of this accountability. It is important, therefore, to consider what action needs to be taken to discharge this responsibility. The risk of fraud and corruption or inefficiency should not be underestimated.

Unless steps are taken to detect and prevent fraud, to critically examine the effectiveness of controls and to deter the potential for fraud, there is a risk. Worse still, it is possible that many organisations are unaware of the risk, preferring to believe they are in control and are getting value-for-money rather than checking that they are. To put things in perspective, the amount of fraud and corruption in maintenance and small works is probably no more than in any other type of business, but still warrants detection and prevention.

Proper control of Measured Term Contracts needs to be carefully thought out, but this should not be seen as detracting from the other benefits and advantages that they have to offer. There is less of a risk when there is a combination of the checking carried out, and when the quantity surveyor is involved. That is not to say that the quantity surveyor is incorruptible, but his professional presence should increase the safeguards.

All of the alternative methods of contracting used in the construction industry have some risk factor attached to them. Jobbing is easy and quick for small value work, but the quotation system is open to abuse and rarely gives the competitiveness of a Measured Term Contract. Daywork term contracts will inevitably suffer from the same risk as any other form of Daywork, while the Small Lump Sum Contract for small works has attracted a good deal of adverse publicity in the past.

The Lump Sum Contract is based on specification, and sometimes drawings, and has been the main alternative to the Measured Term Contract in the public sector for many years. It was intended for use where specialist work was involved or work that could not be covered by the Schedule of Rates, but it has been used by some as the main method of procuring maintenance work.

The competitive lump sum method was a central feature of the fraud trials held at the Old Bailey in the 1980s where the contracts involved were shown to be anything but competitive. Although the irregularities concerned lump sum rather than term contracts, they showed that serious and widespread malpractice can occur within a system of well-established procedures.

During the trials, a number of public employees and building contractors either pleaded, or were found, guilty of corruption and conspiracy to defraud on a series of Lump Sum Contracts in London. The evidence, which was accepted by the four separate juries, showed that, despite apparent genuineness, the following irregularities had taken place:

1. The widespread ringing of contracts had been aided by officials and had involved the sharing of the local workload by a small group of contractors.
2. Contracts were arranged at grossly inflated price levels, often two- or three-times the true value, sometimes 10-times or more.
3. Work frequently did not meet specification requirements and on occasions was not executed at all.
4. Bribery in the form of holidays, houses, furniture, hospitality and large sums of cash was accepted as routine.

The significant aspect is that, despite being grossly overpriced, all of these contracts satisfied the procedures in place at that time:

1. Pre-tender paperwork was invariably in order and an estimated value was always included on the file.
2. Tenders were invited from four or five firms on the Approved List, were received by the due date, and recommended as fair and reasonable.
3. Invariably the tender would be within 10% of the estimate, often very much closer. If not within 10%, an explanation would have been required to satisfy procedures.
4. Work was then apparently executed and signed off as complete.

Anyone looking at the paperwork, (particularly a systems audit) would find contract files that met all the requirements of the system.

A series of measures are required to prevent this type of malpractice. One of the measures is to recognise the merit of an independent inspectorate that would undertake surprise visits, providing both a deterrent and the opportunity for an examination of all contract methods. In addition, it would be able to deal with any investigation that may arise from concern, tip-offs, anonymous letters or any other source.

Risks associated with Measured Term Contracts

Measured Term Contracts do not necessarily attract more fraud than other forms of contracting. Investigations into fraud and corruption frequently relate to maintenance contracts other than Measured Term Contracts; some of the more major frauds are committed on Lump Sum Contracts. However there are a number of risks particularly associated with them. The following are some of the risks involved with Measured Term Contracts:

Bona-fide tenders

The tender for a Measured Term Contract is usually a much more complex offer than a straightforward lump sum tender. It often includes a number of different contract percentages and also rates for Daywork. This makes it more difficult to evaluate tenders submitted by a number of tenderers, each containing varying percentage adjustments to the Schedule of Rates. Tenders need careful and professional evaluation if accusations of manipulation of the process are to be avoided (see Chapter 5).

Approved Lists of contractors are popular and, if carefully and comprehensively put together, provide the assurance of a reputable and financially sound Contractor. This is particularly relevant for Measured Term Contracts where the contract period is three years or more. If the vetting of firms is not professional and thorough or, more significantly, if the list is restricted to a handful of firms who are repeatedly asked to tender against each other, then there is a risk of actually increasing the opportunity for a 'ring' to operate.

The answer is to select carefully, vary the firms, ensure privity of information, and check that tenders are fair and reasonable, represent value-for-money and make use of periodic independent inspection.

Ordering and supervision

As with any form of contract, both Contractor and Employer must know what is to be provided and the level of workmanship that is acceptable. Doubt and misunderstanding will breed conflict and possible termination. Poor ordering is a constant source of problems and also offers scope for malpractice. Good ordering and good supervision are two of the most important ingredients of a successful Measured Term Contract. There has been a tendency to view maintenance work as the poor relation to new work, which is unwarranted as the overall spend is often as high and the need to supervise and to sign-off work as complete is just as relevant.

Record dimensions

It is very tempting to use record or standard dimensions for repetitive work to buildings or areas of land. Painting of housing estates and grass cutting to defence bases are two typical examples. The advantages in saving of resource costs are obvious, but if used it is vital that they are checked at appropriate intervals by the certifying officer and the quantity surveyor to ensure that no revisions have taken place.

Record dimensions being used for quite a few years where they are no longer applicable can lead to needless and serious overpayment. Any error is repetitive. For example, in one case, some record dimensions for painting were examined and it was found that one dimension (for the floor to ceiling height) had been carefully changed from 4 metres (13 feet) to 40 metres (130 feet) which had led to the cost of painting a house being accepted at an inflated level. In other words, not only was the error repeated year after year, but because the spend was virtually the same as the previous year no-one was concerned.

When record dimensions are used, be assured of their accuracy.

Agreed Rates and Daywork

Where work cannot be based upon the Schedule of Rates, the contract provides for the use of Agreed rates and Daywork. Without any implied criticism of either the quantity surveyor or the Contractor, this is an area that inevitably attracts close scrutiny. Interpretation of the

Schedule of Rates itself is also often the subject of disagreement and should also merit close attention.

Quotations are another problem area. How often are they accepted on face value? How often are discounts and trade prices accepted, and how does the quantity surveyor know whether there is an arrangement for end-of-year credit between supplier and Contractor?

Daywork, too, can be a problem and, where it is unavoidable, needs firm control. Apart from the inflation of hours, the overpricing of materials and the lack of approval for daywork, one of the most serious abuses has been in duplication. Not necessarily the same men being recorded at the same time on the one Measured Term Contract, but sometimes on a number of contracts at nearby locations. Forty-two hours in a day for one man is very hard to believe.

There is also the potential for duplication between measured work and Daywork.

Duplication

Even the most searching audit can fail to spot blatant fraud in this area.

For example, a gymnasium is to be refurbished at a cost of about £15,000. An Order is raised on the Measured Term Contract and the work executed. Subsequently, the Order is selected for audit and the quantity surveyor duly checks and finds that the work done and the account correctly priced. There may be no obvious indication of malpractice as, understandably, the quantity surveyor has only looked at the particular Order in front of him.

However, an unscrupulous person could have raised more than one Order for the same work, with a slightly different description or location. Such an attempt would normally be spotted, but suppose a different Contracting method was used, such as a fictitious Lump Sum Contract, or jobbing order? Whichever contract is checked, the work will always appear to have been executed.

This may seem an unlikely occurrence, but it has happened, especially when a variety of procurement methods are available.

Non-execution of work

This revolves around the responsibility of the ordering officer to sign off work as being complete. It is likely that he cannot see everything and it may be necessary to decide whether he should be allowed to certify work as being complete but unseen by him. It is also possible that certain work gets covered up (as in decoration) or that one grass cut becomes indistinguishable from the next. These areas need consideration, but there is very little to prevent a corrupt certifying officer from signing off work as complete when he knows it has not been done, especially if he knows it will be almost impossible to check at a later stage, as in the case of drain-clearing or grass-cutting.

Audit procedures

Aims

When auditing Measured Term Contracts, the user is seeking a measure of assurance that the accounts accurately reflect the value of work properly ordered and executed; in other words, that they are in accordance with the contract. The factors to be taken into account are:

- assurance
- deterrence
- value for money
- quality of work executed, and
- cost-effectiveness.

Taking each of these factors in turn:

Assurance
There is a need to confirm that Measured Term Contract accounts are within tolerable levels of accuracy. The aim is to advise, and hopefully to assure, the Employer on the level of accuracy of MTC accounts, so that action can be taken if there is any apparent cause for concern. Absolute assurance can only be gained if a complete check is made on the account of each Order of work raised. Clearly, an approach such as this represents an element of over-kill.

Sufficient information for the purposes of keeping the Employer informed can be obtained from the results of checking a statistical random sample of accounts. The overall number of orders placed on a Measured Term Contract will directly affect the size of the random sample and statistical advice may be needed on this. It does not necessarily need to be a large sample to be sufficient. In fact, quite a small sample may suffice.

Some important points to bear in mind are:

1. For statistical reliability, the results must come from a 'Full Technical Check' of each Order identified by the sample. This means fully re-measuring, on site, and checking the pricing of all items on each sampled order.

2. The sampling process must be unknown to the person whose work is being checked. This is rather obvious, but if it is known which accounts are to be checked it is highly probable that those accounts will not be found wanting. This could distort the results and disguise the real underlying trend.

Deterrence
The aim is to seek to deter those who are responsible for measuring on Measured Term Contracts from misrepresenting the amount properly due on any account, either through inefficiency or with intent to defraud. It goes without saying that any form of random checking has some inherent amount of in-built deterrence. But whereas a statistical sample with known limits of reliability can be devised for information purposes, it is not possible to measure deterrence in statistical terms. If all orders were checked, and known to be checked, perhaps deterrence could be assumed to be absolute and nobody would try to defraud the system, but this could not be relied upon. In order to establish deterrence, a system needs:

- Publicity: it must be widely known that checking is carried out. People will not be deterred if there is no risk of being found out.
- Confidentiality: the orders to be checked must be unknown to those people who measure them and the size of the sample is best kept confidential.
- Action: firm action will need to be taken against those found to be in serious default and potential defaulters must be made aware of the consequences of their actions.

Value for money
The user must be sure that Measured Term Contracts are being appropriately used. Other methods of procurement, usually lump sum tendering, may sometimes be more

appropriate and should not be ignored. For example, if an Order for an extension to an existing building valued at £100,000 is issued against a Measured Term Contract for minor repairs to housing, it could be regarded as an inappropriate method of procurement. On the other hand, the Contractor will rightly expect a fair balance of workload for the Measured Term Contract to remain viable to him.

Quality of work executed
A Measured Term Contract account should reflect only the quality of work actually provided by the Contractor, which could sometimes be inferior to that ordered. For example, a Contractor may deliberately use materials of lower quality or smaller quantity than specified on the Order, or carry out work of poor workmanship to ensure further repairs are necessary in the future and guarantee him more work. Suitable reductions to the value of the Order would be required. The alternative would be to instruct the Contractor to rectify the deficiency. Conversely, the Employer will not normally expect to pay for better quality than was specified or for more work than was ordered.

In the Measured Term Contract, auditors should be required to take account of quality of work in the remeasured value of an audited account.

Cost-effectiveness
Experience has shown that a properly structured audit procedure will result in a reduction of overpayments on Measured Term Contract accounts. The statistics derived from analysis of the various checks undertaken show that the amount spent on audit is less than the total overpayment that would otherwise be incurred.

But how far is it worth chasing the error rate down in search of further savings? It would be fair to say that a publicly accountable Employer might take a different attitude to an Employer from the commercial sector. Further cost-effectiveness could result from the statistical analysis by identifying where additional checking could best be directed to obtain the greater recovery of overpayments. The use of risk analysis should be given consideration.

Matters for consideration

When formulating an audit procedure, a number of questions will need to be resolved, some of which have already been touched upon above. For example:

How many checks should be carried out?
This will depend on a number of factors, such as the total number of Measured Term Contract accounts being passed for payment, and what assessment is made of the risk.

What amount of detail should a check comprise?
Real information can only come from results of Full Technical Checks, but in addition some form of brief or spot checks may also be appropriate.

When should the checks be carried out?
Checks carried out prior to payment could affect the Contractor's cash flow, but on the other hand any overcharging (or undercharging) can be immediately rectified and only the amended account need be paid.

Who should carry out the checks?
The obvious choice is that the checks should be shared between in-house personnel and outside consultants. If in-house personnel are used, individuals carrying out inspection duties must be separately accountable under their job descriptions for their activities as auditors. If consultants are employed, assurance will be needed that they are able to

assemble teams of appropriately qualified professional staff to conduct audits. The staff used should not only be experienced professionals in their particular discipline, but also have been involved in similar work.

What action should be taken when accounts are found incorrect?
Only firm action will be a deterrent against inaccurate accounts. 'Firm action' will need to be defined and be appropriate to the accountability of the Employer, principally, in terms of whether he is public or private sector, to the size of the inaccuracies and the frequency of occurrence.

Risk evaluation

In order to focus resources on the key areas susceptible to fraudulent activity, a risk evaluation should be carried out as an integral part of the audit process. In broad terms, the process involves identifying functions that may be the subject of fraudulent activity, e.g., over-measurement of works, collusion with suppliers, etc. By evaluating the probability of malpractice occurring in these functions along with the financial impact, a weighted-value can be calculated with the result that a prioritised list can be produced. This prioritised list would provide a means of directing the auditor to specific functions covered by the audit.

Reviews and audits

The requirements of audit procedure tend to vary according to whether the organisation is in the public sector, where they are publicly accountable for all the money that they spend, or the private sector, where they are usually accountable to shareholders. The principles are the same, but, as public accountability is more sensitive, the following focuses more on public sector requirements.

The audit of Measured Term Contracts generally consists of:

1. reviews of procedures; and
2. financial audit of accounts.

Reviews of procedures
Specialist teams regularly review and report to the Employer on the work of each local office or region. The contract strategy and whole operation, of which Measured Term Contracts form only a part, are critically examined. In particular, the review will take account of value-for-money and quality of work achieved. This will include on-site inspections of a selection of work executed, including some Measured Term Contract Orders. Additionally, the senior manager at each location is responsible for carrying out his own checks on a selection of work, chosen at random, to satisfy himself on the effectiveness of those who work for him.

Financial audit of accounts
The major effort should be concentrated here. In the day-to-day management of all aspects of the Measured Term Contract, managers need to be aware of the type of indicators that should be monitored in order to provide an early awareness of the possible presence of fraudulent activity. The presence of fraud can be detected early if the following indicators are regularly noted and monitored by managers:

- incorrect authorisation of orders and invoices (must be separate)
- payment for services not authorised by a manager senior to the person ordering the service

- inadequacy of specifications for procurement
- high prices for similar work – value-for-money checks
- work not to specification
- poor quality work
- Contractor patronage
- use of photocopied invoices for payment
- evidence of altered records
- incomplete tender records
- lack of proper sign-off to confirm that the services have been completed to specification and cost.

In addition to these work-related indicators, there are other 'personal' trends that should be regarded as indicating the possibility of dishonest or illegal activity. These include:

- the employee is the only one fully trained in the use of the systems (IT)
- over elaborate processes initiated by the individual to deter third-party involvement and deter thorough audits
- regularly works late in the office
- frequently works weekends
- rarely takes all leave entitlement.

In the vast majority of cases, fraudulent activity is uncovered by chance and it is probable that any combination of these items could indicate that the employee is 'covering his tracks' and ensuring that few other people have access to his paperwork and records. When the employee is absent, then other staff trying to access the information could accidentally uncover malpractice and expose the fraud.

Strategy

Within an overall percentage number of accounts to be checked each year, procedures should be designed to provide for:

- Audit information obtained from the results of checking a statistical random sample of accounts across the whole of the Employer's organisation.
- Local management selection of some accounts to be checked.
- A minimum number of accounts to be checked annually on each Measured Term Contract. This is usually in the region of 5%.
- A greater concentration of checking on Measured Term Contracts on which the error rate is found to be unsatisfactory.
- A greater concentration of checking on Contractor self-measured accounts.

Features

Principal features of the procedures should be:

- All checks must be Full Technical Checks.
- All checks are carried out post-payment. Due consideration should be given to the Contractor's cash flow, except where there is particular cause for concern.
- More checks than the prescribed minimum are carried out if circumstances dictate and particularly where there is any cause for concern.

- Delegation to local managers to ensure successful operation of the procedures and to be responsible for the level of accuracy of Measured Term Contract accounts in their area of responsibility.

Selection

The selection of accounts to be checked is divided (but not equally) into three distinct parts:

1. *Statistical random sample:* this provides information on the overall level of accuracy of Measured Term Contract accounts. An established method is to use a random numbers programme in the computer that makes the actual payments of MTC accounts. There is no way of knowing which accounts will be checked because an account only becomes liable for selection once it is logged on to the computer for payment.

2. *Local management selection:* the statistical random sample is not designed to achieve the overall amount of checking required each year; the additional accounts required are selected by local managers. They will use their discretion based on local circumstances, giving more attention to Measured Term Contracts where the error rate is greatest. They will also ensure that a prescribed minimum amount of checking is carried out on each MTC for which they are responsible.

3. *Additional checking on unsatisfactory Measured Term Contracts (continuous checking):* a set of error limits should be established, on a sliding scale related to value of order, beyond which an account is deemed 'unsatisfactory'. This is the trigger for additional checking to be carried out on that particular Measured Term Contract. The objective is to discover whether or not an unsatisfactory account was just a rogue, or part of a trend. Continuous checking is therefore carried out on that MTC until either confidence is restored or it is decided to take more drastic action, which is usually found to be unnecessary.

The selection procedure should also require that:

- One (or more) accounts is checked in a given period from orders raised by every person who orders work on Measured Term Contracts.
- Similarly, every person who measures on Measured Term Contracts should be so checked.
- Where consultants are employed to undertake checking of Measured Term Contract accounts, some of the accounts checked by them are re-checked by the appropriate manager within the Employer's organisation to ensure that quality of checking is of the required standard.

Dealing with errors

Tolerable limits of accuracy should be established and any overpayments discovered outside those limits are recovered from the Contractor. Likewise, underpayments are reimbursed. It is not usually necessary for Contractors to make separate reimbursement of overpayments, nor for the Employer to pursue recovery. Standard Contract Conditions usually provide for recovery to be made from any monies still owed to the Contractor. For example, Condition 34 (Recovery of sums) of GC/Works/7 gives the Employer additional rights to deduct any sums recoverable from sums due under the Contract or any other contract between the Contractor and the Employer.

Local responsibility

Complying with the prescribed procedures laid down by the Employer is not enough for local managers to discharge their responsibility. They should be required to do whatever is considered necessary to satisfy themselves that Measured Term Contracts within their area of responsibility are under control. The level of checking undertaken should reflect local circumstances, and checking over and above the prescribed minimum may sometimes be necessary.

As a supplement to the prescribed procedures that require only Full Technical Checks, some local managers may install either complete or partial systems of 'brief' or 'spot' checks on accounts prior to payment, usually referred to as 'desk checks'. The results of 'desk checks' are often used to decide which accounts should form the basis of locally selected accounts for Full Technical Checks.

Feedback

The procedures should provide for regular feedback of information to the Employer and his local managers.

In addition to receiving a detailed analysis of the statistics obtained from the results of the statistically random sample, feedback is also given on who are the persistent offenders in preparing unsatisfactory accounts. The Employer is automatically made aware of any cause for concern.

The procedures, therefore, are designed to ensure that the Employer is sufficiently informed to make judgements on the operation of Measured Term Contracts throughout the his organisation and take any further action deemed necessary to ensure they are under proper control.

Conclusion

In conclusion, it is fair to point out that practically every action that is taken during the management of a Measured Term Contract is open to abuse, but as this is also the case with other forms of contract there is no cause for particular alarm. The Measured Term Contract is as safe, if not safer, than most and will, if used properly, provide the best, balanced, approach to maintenance needs.

12 The Contractor's view

Introduction

This chapter gives the view of the Measured Term Contract from the perspective of the Contractor. Measured Term Contracts provide Contractors with profitable, regular work where there is sufficient workload to offer continuity and economy. It is important that all work covered by the Schedule of Rates is ordered from the Contractor, and not just the low-value and awkward work. The Contractor will have tendered on the basis of all relevant work with a spread of value and will be investing capital to provide the resources needed over the contract period.

The Measured Term Contract has, however, distinct disadvantages for the Contractor over 'cost-plus'- type contracts in that the price paid to the Contractor is defined in a Schedule of Rates and consequently, the Contractor carries the responsibilities of his own inefficiencies. He also needs to be in a position to respond quickly to normal demands and to be able to deal with any emergency work that is needed.

Estimating and tendering

Main factors

There are a number of key factors that will influence the Contractor, including:

Location
The location of the site is an important factor in the Contractor's overhead costs. A site covering a large area will affect the time spent travelling to Order locations and the operational cost of vehicles.

Accessibility
Repairs and maintenance works to flats and houses often involve a large amount of time spent making appointments to gain access to individual properties. Large buildings on high-security military bases may be easier to gain access to, but there may be additional costs in security clearance or waiting time to enter the establishment at the beginning of the day. The JCT Standard Form 1998 states that, unless it is otherwise provided for by the Schedule of Rates, the Contract Administrator is responsible for arranging access to the Site. In practice, Employers frequently prefer the Contractor to arrange for his own access, particularly when working in occupied premises.

Contract period
The duration of the contract and the associated problem of how to allow for fluctuations over the term are important considerations. Contracts can have a contract period of one, two, three or even five years. One-year contracts are uncommon as the set-up costs for the Contractor can be high; only after three to five years is he gaining a significant advantage from continuity. However, if the contract period is of this duration, there should be a

fluctuation price condition in the contract. The Contractor cannot reasonably be expected to predict inflation over this time period. It is common practice for the Contractor to price the first year as a fixed price. If required to price a two-year term contract fixed price, he may price for the first year, when there will be minimal inflation, and then exercise the break clause to terminate the contract to avoid working in the second year at a lower rate.

Type of buildings

The type of building should also be considered. For example, in the case of public sector housing estates in London, the high-rise buildings could be kept separate from low-rise buildings, or the Contractor could be given the opportunity of quoting a different price. Many factors affect the pricing. The renewal of a bath on the twelfth floor of a block where the lifts do not work requiring it to be carried manually up the stairs warrants a very different kind of price to replacing the same bath in a low-rise property. It is important that the Employer thinks very carefully about the packaging of buildings and areas in the tender.

Pricing

When invited to tender for a Measured Term Contract, the Contractor must consider how he is to arrive at his tendered price. The process is relatively straightforward when pricing a Lump Sum Contract. The work is defined and the Contractor can examine the location, define the scope and quantify the work by measuring on site or from drawings.

A Measured Term Contract is an arrangement whereby the Contractor will undertake to carry out a series of Works Orders, over a period of years, within a defined geographical area and where the work is subsequently measured and valued at rates contained in a pre-priced Schedule of Rates. At the time of tender, he is unlikely to know the content of these future Works Orders. This, obviously, makes tendering much more difficult and risky. It is important that the Employer supplies tenderers with as much information as possible about the estimated annual value of the work to be ordered, the type of work and the size of orders. The following is one way that a Contractor can arrive at a percentage addition to the Schedule of Rates:

Table 12.1 shows a comparison between typical rates from Bills of Quantities and the PSA Schedule of Rates for Building Works (Eighth Edition). A selection of rates used for pricing a Bill of Quantities, built up in the normal method, are compared with the prices in the Schedule of Rates for the same type of work. The Bill of Quantities rates must be the actual rates paid including overheads and profit. If overheads and profit are shown separately at the end of the Bill of Quantities, the rates will need to be adjusted. A Contractor estimating against a Schedule for the first time should take at least 15 rates in each section of a Schedule, and preferably more repair items, and compare them with what he would normally charge.

Table 12.1   Comparison of rates

Item	Description	Schedule of Rates (£)	Bill of Quantities (£)
D20.014/1	Excavate trench for foundations	7.45	8.20
E10.005/1	Foundations 150 to 300mm thick	86.14	83.56
F10.001/1	Half brick wall in gauged mortar	21.22	23.98
F30.444	Galvanised lintel 1050mm long	21.00	22.26
G20.036/1	Sawn softwood 50 x 100mm	3.24	2.44
J41.145/1,4&5	Three-layer built-up roofing	25.21	15.13
K20.382/3	Repair 21mm T&G floor	27.99	19.56
L40.384/2	4mm obscure glass 0.50 to 4.00 m^2	35.82	40.12
M20.194/3	Two coat lightweight plaster	7.75	6.74
M20.269	Cut out and make good plaster crack not exceeding 50mm wide	3.06	2.89
M60.657/1&2	Prepare, prime and two coats oil paint to wood	3.91	4.38
N13.148	Re-washer tap	4.39	3.75
Q40.189/1	1200mm high close boarded fence	42.07	24.82
R12.453/1	100mm clay drain, flexible joints	8.47	6.86
S10.107/2	Copper tubing 22mm to softwood	6.17	7.59

The Contractor must now estimate the frequency that various types of work are likely to occur. Table 12.2, below, shows a possible weighting of work categories. This can be produced by assessing that out of a total of 300 work elements of equal value, there will be, for example, only four work elements in Excavation and Earthworks on this kind of maintenance contract and 150 plaster repairs.

Table 12.2   Possible weighting of work categories

Excavation and Earthwork	4
Concrete Work	9
Brick and Block walling	13
Accessories to brick and block	10
Carpentry timber framing	20
Built-up felt roofing	4
Timber board flooring	17
General glazing	10
Plastered rendered and roughcast coatings	13
Plaster repairs	150
Painting and clear finishing	10
Sanitary appliances and fittings	5
Fencing	5
Drainage below ground	10
Cold water	20
Total	300

The Contractor needs to be reasonably confident of the 'mix' of work he will expect to get. The proportions between new work and repairs, and the spread of work over the various sections of the Schedule of Rates are all important factors. It is in his interest to discover these from the Employer at tender stage. Unfortunately, in practice, Employers rarely supply this and the Contractor will need to make his own assessment.

Table 12.3 shows that by applying the comparison rates from Table 12.1 to the weightings in Table 12.2 an overall nominal total can be produced. The example shows that the Schedule of Rates prices would give a 7.75% average increase over the Contractor's usual charges. This is based purely on costs without any other factors such as location, volume, economy of scale, etc.

Table 12.3   Possible totals

Item	Description	Schedule of Rates	Bill of Quantities	Weighting	Schedule of Rates	Bill of Quantities
		£	£		Total £	Total £
D20.014/1	Excavate trench for foundations	7.45	8.20	4	29.80	32.78
E10.005/1	Foundations 150 to 300mm thick	86.14	83.56	9	775.26	752.00
F10.001/1	Half brick wall in gauged mortar	21.22	23.98	13	275.86	311.72
F30.444	Galvanised lintel 1050mm long	21.00	22.26	10	210.00	222.60
G20.036/1	Sawn softwood 50 × 100mm	3.24	2.44	20	64.70	48.80
J41.145/1,4&5	Three-layer built-up roofing	25.21	15.13	4	100.84	60.50
K20.382/3	Repair 21mm T&G floor	27.99	19.56	17	475.83	332.52
L40.384/2	4mm obscure glass 0.50 to 4.00 m^2	35.82	40.12	10	358.20	401.18
M20.194/3	Two coat lightweight plaster	7.75	6.74	13	100.75	87.65
M20.269	Cut out and make good plaster crack not exceeding 50mm wide	3.06	2.89	150	459.00	433.50
M60.657/1&2	Prepare, prime and two coats oil paint to wood	3.91	4.38	10	39.10	43.79
N13.148	Re-washer tap	4.39	3.75	5	21.95	18.75
Q40.189/1	1200mm high close boarded fence	42.07	24.82	5	210.35	124.11
R12.453/1	100mm clay drain, flexible joints	8.47	6.86	10	84.70	68.61
S10.107/2	Copper tubing 22mm to softwood	6.17	7.59	20	123.40	151.78
Totals					**3,329.74**	**3,090.30**

Note: Amount by which SoR rates are to be reduced to reach BoQ level = 7.75%

Sub-contracting

Many Contractors sub-contract work, particularly trades such as plumbing or felt-roofing. Consideration must be given to the Contract Conditions with regard to sub-contracting. Some Employers are very strict about how much work they will permit to be sub-contracted, believing that there is a loss of control when the Contractor's directly employed workpeople are not employed.

Preliminaries

Table 12.4, below, shows the build-up of preliminaries for a Measured Term Contract of £1,000,000 estimated annual value (EAV). It assumes that the Site cost would be 85% of that £1,000,000.

This produces a cost of £850,000 per year. Assuming 20% is going to be sub-contracted, this gives a cost for the main Contractor's work of £680,000. Assuming labour would be 50% of that cost, because Measured Term Contracts are labour-intensive, and working on a cost of £20,000 per man, gives a required workforce of 17 men.

£1,000,000 x 85%   850,000

Less
Value to be subcontracted 20% = <u>170,000</u>
                                   680,000
Labour = 50%   340,000

Labour = 340,000 @ £20,000 per man = 17 men

A Measured Term Contract with over ten men working may require a non-working foreman or site agent. Ten men or under may require a working foreman who can spend some of his time 'on the tools', but this can vary according to circumstances and individual Employer's requirements. This example shows a preliminary build-up for the first year of the contract.

Table 12.4   Possible preliminaries, build-up

Preliminaries	£
Site agent or general foreman	30,000
Car or other personal transport	4,000
Site Clerk/Liaison Officer	15,000
Site accommodation (heat, light, etc.)	4,500
Telephone	2,300
Welfare facilities	4,500
Rubbish removal	2,500
Disposable plant	3,000
Transport	16,000
Sub-total	<u>81,800</u>
Sundries 10%	8,180
Total	<u>89,980</u>
Say	**90,000**

Tendered percentages

The next stage is for the Contractor to calculate a percentage adjustment to the Schedule of Rates to insert in his tender to the Employer. Table 12.5, below, shows a typical tender percentage calculation and is based on information in Tables 12.1 to 12.4. It shows that the cost of work using the Contractor's normal Bills of Quantities rates would be £850,000 and that the site preliminaries are £90,000 (see Table 12.4, above) giving a total of £940,000. The same work valued at Schedule of Rates prices would be £915,875 (£850,000 plus 7.75%). The difference between the total value of the works, £940,000, and the value based on the SoR, £915,875, is therefore £24,125. Expressing this as a percentage (£24,125 divided by £915,875) gives a net addition of 2.63%. This is the tender percentage that will, hopefully, secure the work and a small profit, subject to any further adjustments he may wish to make to allow for overheads, profit or financing charges.

Table 12.5   Percentage calculation

Cost of work (less preliminaries)	£850,000	
Value at SoR (£850,000 + 7.75%)		£915,875
Add preliminaries	£90,000	£24,125
Original Total	£940,000	£940,000
'Tender' Percentage Addition on SoR (£24,125/£915,875)		**2.63%**

Overheads and Profit and Financing Charges

Overheads on a Measured Term Contract generally work out higher than they do on a Lump Sum Contract as the measuring and valuing element is more expensive, possibly 4% or 5% of total costs. The Contract should specify whether the Employer's representative or the Contractor is responsible for measuring and valuing. However, regardless of what the contract states, the Contractor needs to measure it, if only to check that nothing is being missed.

The cost of financing the contract can be very expensive. The Contractor only has to look at contract conditions which state 56 days in which to measure, 28 days to check or certify and 14 for payment. Even if he is quite efficient, he may have to carry out several month's work ahead of getting paid for it.

Whether liquidated damages are included in the contract or not, there is an incentive for the Contractor to get the work done as quickly as he can. He will want to get the work completed, measured and valued as, until he completes the entire process, he will not get paid. Most contracts have provision for progress payments but these tend to be on larger Orders. In practice, most are not large, and merely reflect day-to-day maintenance and minor repairs. Large numbers of these small Orders will soon represent a considerable financial burden for the Contractor.

The reasons for delays in completion are varied but they are not always within the Contractor's control. For example, it is often necessary to match existing materials and as a result delivery times can be slow. Occasionally, a particular material may no longer be available and the Contractor must consult the Employer's representative on a suitable alternative, which may then also be subject to delays. It is not always a question of the Contractor taking his men off the contract to carry out more profitable jobs; it is very often just a question of delay in getting the right kind of materials or a lack of pre-planning.

Pre-planning is an aspect that is rarely given sufficient importance in the administration of a contract. A Measured Term Contract will frequently contain a fair proportion of repair work, which, by its very nature, tends to be urgent. The Contractor may receive an Order that was prepared some time after the damage had been reported, but the Employer may still expect the work to be completed immediately. The Contractor will frequently be blamed for a month's delay before he even starts because not all the facts are known. Delays are not always the Contractor's fault. A good working relationship is essential. This is becoming more recognised with the increasing tendency for Employers to adopt 'partnering' principles and the development of a long-term relationship with a Contractor.

Addition for fixed price

Calculation of the addition for the fixed price if required on the first year of a contract is shown in Table 12.6, below. This shows a fixed price adjustment to the contract in the other examples, but the formula would apply to any contract of a year's duration with adjustment of the labour and materials relationship as applicable. The resultant 2.54% is applied to get the final percentage adjustment.

Table 12.6   Addition for fixed price

Take a 12-month contract start date: 1st April. Contract will run for 12 months but allow an extra 2 months for completion of outstanding work giving a total of 14 months.

Labour	= 50% of cost)

Assume 5% wage increase from 1st July

First 3 months = 0%

Next 11 months = 5%

$\dfrac{11 \times 5\%}{14}$ $=\dfrac{3.93\%}{2}$ = 1.96%

Materials and other costs   (= 50% of cost)

Inflation rate annually   = 2%

Take $\dfrac{14}{12} \times \dfrac{2\%}{2}$ $=\dfrac{1.17\%}{2}$ = 0.58%

To add   2.54%

The Contractor's final Tender Percentage adjustment would therefore be an addition of 2.63% (Table 12.5) on the Schedule of Rates and a further addition of 2.54% for Fixed Price, i.e., 1.0263 x 1.0254 = 1.0524, or **5.24%**.

13 Summary

We have seen that the Measured Term Contract is one of the methods of obtaining value for money in maintenance and has distinct advantages over 'cost-plus'-type contracts in that the price paid to the Contractor is defined in a Schedule of Rates.

The future

The use of Measured Term Contracts has increased rapidly in recent years as government initiatives have led local authorities and others to introduce competitive tendering for maintenance work, and the Measured Term Contract has become the cornerstone of the maintenance operation. The development of new initiatives is continuing with the introduction of Partnering, Prime Contracting, Public-private partnerships and the Private Finance Initiative; these are discussed in turn below.

Partnering

There is current trend to move away from the adversarial nature of construction contracts in all aspects of construction. Competitive tendering has always been seen as the solution to the search for value-for-money, but, increasingly, the industry is coming to realise that there is much to be gained from long-term relationships between the Employer and the Contractor, and the Contractor and his Suppliers. Developing long-term relationships and this associated 'supply chain' is essential to implementing the concept of Partnering. To regular users of Measured Term Contracts, however, there is nothing new about this concept. Measured Term Contracts have been delivering the benefits of Partnering for many years. Over the term of the contract, the Employer benefits from the Contractor's increasing familiarity with the buildings while the Contractor benefits from the assurance of a reasonable continuity of work. There is an excellent opportunity for both parties to create a good long-term working relationship

Prime Contracting

Prime Contracting is 'single point' responsibility integrating the supply chain and effective project management. The advantage of having one Prime Contractor is that direct long-term relationships with suppliers can be established delivering improved quality. Adversarial relationships are reduced and mutual support is increased. Economies in construction activities deliver better value-for-money for the Employer and improved profitability for the Contractor.

Public-Private Partnership (PPP)

Partnership between the public and private sectors is part of the Government's strategy to modernise key public services. Public-private partnerships cover a range of business structures and partnership arrangements, such as the Private Finance Initiative (PFI), joint ventures, outsourcing and the sale of equity stakes in state-owned businesses.

Private Finance Initiative (PFI)

The Government is seeking value-for-money from involving private sector management expertise, innovation and capital investment in the delivery of services to the public sector. PFI is a procurement approach aimed at determining the feasibility and cost-effectiveness of allowing the private sector to provide the public sector with certain capabilities and services. PFI differs from privatisation in that the public sector organisation retains the ultimate responsibility for ensuring service delivery. The private sector is required to invest in, manage and operate any capital assets necessary to deliver the service. The public sector as the customer of the service pays directly through a long-term contract, frequently 20 to 30 years.

A new role for the Measured Term Contract

Initiatives come and go but the concept of measuring and valuing repairs and minor works against a Schedule of Rates has stood the test of time and will be with us for many more years to come. The future may be with the Prime Contractor as 'Employer' but the Measured Term Contract will always be needed when it comes to delivering value for money in maintenance.

In conclusion, the benefits of using Measured Term Contracts based on a Schedule of Rates include:

- *Demonstrable value for money* – value for money is demonstrable as the contract percentages used to value work are the result of a competitive tendering process that involves comparative checks. The continuity and volume of work should also result in more competitive rates and consequent cost savings for the Employer.

- *A highly regarded reputation* – they are a well-tried and tested method of successfully procuring maintenance and new works, both pre-planned and responsive.

- *Flexibility* – there is no more flexible method of procuring an ongoing series of small works, while retaining public or shareholder accountability.

- *Easy ordering* – a multitude of Works Orders can be raised under one contract with very little documentation.

- *Speed of response* – the Contractor's response should be immediate once an Order is issued, as new contract action is not required. Minimum and maximum response times can be specified in the contract documentation.

- *An effective basis for emergency work* – apart from being in a position to respond quickly to normal demands, the Contractor should also be able to deal immediately with any emergency work that is required.

- *Time and cost savings* – savings in time and resource costs should result from this type of contract documentation.

- *Continuity* – the Employer benefits from the Contractor's increasing familiarity with the buildings, while the Contractor benefits from assurance of a reasonable continuity of work.

- *Establishment of good, long-term working relationships* – there is an excellent opportunity for both parties to create a good, long-term working relationship under independent scrutiny.

- *Documentation is easy to produce and operate* – tender documentation, application of percentage adjustments and ordering work are all simple operations.

- *Auditability* – accountability, probity and value for money can be demonstrated through cost effective auditing.
- *Scope for computerisation* – the ordering, estimating, measurement and valuation of works lend themselves to computer applications for ease of administration and control.

Following on from this, the benefits that derive from using published standard Schedules of Rates include:

- *Well-established reputation* – current editions represent a long-term development and are widely accepted by the industry.
- *Readily available* – as published documents, they are already commercially available and can, therefore, be used immediately.
- *Specification standards are defined* – this has the effect of reducing the amount of resource time required when ordering work.
- *Flexibility* – they are flexible in use and can be used for one-off contracts as well as Measured Term Contracts.
- *Benchmarks* – they provide a benchmark for comparing regional variations and monitoring tendering trends.
- *Ease of use* – they are easy to use and understand, and are presented in an industry standard format.
- *Regularly updated* – new editions are produced to reflect changes in prices, legislation and standards.

MODEL FORM 1
ABSTRACT OF PARTICULARS AND ADDENDUM

Works:

Site:

Condition 1(1) (Definitions, etc.) Employer

The Employer shall be

of

Conditions 1(1) (Definitions, etc): Project Manager, and 3(1) (Delegations and representatives)

The Project Manager shall be

*of/whose registered office is at

who shall act generally on behalf of the Employer in carrying out those duties described in the Contract, subject to the following excluded matters:

In relation to such excluded matters, the person or persons authorised to act for the Employer are:

Condition 1(1) (Definitions, etc): Contract Period

The contract shall run for a period of   years from the commencement of the Contract.

Condition 1(1) (Definitions, etc.): Contract Area

The Contract Area is contained within the following boundaries [insert details] and marked on drawing number [insert details]

Condition 1(1) (Definitions, etc.): (Maintenance Period)

The Maintenance Period for building work is *[insert period e.g. 6 months]*

The Maintenance Period for mechanical and electrical work is *[insert period e.g. 12 months]*

Condition 1(1) (Definitions, etc.): Orders

The minimum value for any one Order shall be £
The maximum value for any one Order shall be £

Condition 27 (Valuation)

The percentage addition under Condition 27 (2) is   %
The percentage addition under Condition 27 (3) is   %
The percentage addition under Condition 27 (4) is   %

Condition 29 (Advances on account)

Under Condition 29 (Advances on account), the Contractor will be entitled to advances on account if the PM estimates that the value of the Works in relation to a single Order exceeds the sum of *[insert value]*. Such sum to be exclusive of VAT.

**Condition 39 (Adjudication)
The adjudicator shall be

of

or, if he is unable to act, or is not or ceases to be independent of the Employer, the Contractor and the PM,

of

or, if he is unable to act, or is not or ceases to be independent of the Employer, the Contractor and the PM; such other person as the Employer and the Contractor choose by mutual agreement in writing or, failing such agreement, such other person as may be chose by the President or a Vice President of the Chartered Institute of Arbitrators (or, where the Contract is a Scottish contract, by the Chairman or a Vice Chairman of the Chartered Institute of Arbitrators (Arbiters) (Scottish Branch)) at the request of either the Employer or the Contractor.

The prescribed form of adjudicator's appointment is appended.

***Supplementary Conditions and Annexes

The following Supplementary Conditions and Annexes (if any) are incorporated into the Conditions of Contract, and shall prevail over the other Conditions of Contract:

*Delete inapplicable items.

**The same adjudicators and aribtrators should be named in all the Employer's contracts relating to the project, whether with contractors, consultants or others.

***It is recommended that any printed Conditions affected by Supplementary Conditions should be amended and initialled by both parties.

MODEL FORM 2
INVITATION TO TENDER

Works

Site:

1   ('the Employer') invites you to tender, upon the basis of GC/Works/7 General Conditions (1999), for the Works described in the following enclosed documents:

    (a)   Abstract of Particulars and Addendum;

    (b)   Supplementary Conditions and Annexes (if any) referred to in the Abstract of Particulars;

    (c)   Specification;

    (d)   Outline Health and Safety Plan; and

    (e)   Other documents as listed below:

2   Your tender should be submitted on the form of Tender also enclosed.

3   You are required to keep you tender confidential and not divulge to anyone, even approximately, what your tender price is or will be. The sole exception to this is information you may have to give to your insurance company, or broker, in order to compile your tender, but you must stress to them that this information is given in strict confidence.

4   You must not make any arrangements with anyone else about whether or not they should tender, or about their or your tender prices or terms and conditions. You may however, obtain any necessary subcontract quotations.

5   No tendering expenses will be reimbursed by the Employer.

6   The Employer does not bind himself to accept the lowest, or any, tender.

7   Your form of Tender should be submitted in a sealed envelope prominently marked:

FORM OF TENDER FOR
WORKS:
SITE:

The envelopes should bear no external indication of the identity of the tenderer.

8 Tenders must be completed and returned by   a.m./p.m. on
       to:

SIGNED by

for and on behalf of the Employer

Tel:

Fax:

Telex:

Date:

MODEL FORM 3
TENDER FORM

TENDER

Works:

Site:

To be returned by   a.m./p.m. on   to

of

1   We have examined GC/Works/7 General Conditions (1999), and the following documents:

(a)   Abstract of Particulars;

(b)   Supplementary Conditions and Annexes (if any) referred to in the Abstract of Particulars;

(c)   Specification;

(d)   Outline Health and Safety Plan (and confirm that we will provide a statement and details of how we plan to implement and develop it, together with details to establish our competence and resources to comply with the requirements and prohibitions imposed upon us relative to health and safety in the execution and/or management of the Works);

(e)   The Schedule of Rates for   *Works and referred to as the 'Schedule of Rates';

(f)   The Updating Percentage Adjustments for Measured Term Contracts giving the Updating Percentage Adjustments to the Schedule of Rates up to and including the month of *   ;

(g)   The Definition of Prime Cost Of Daywork carried out under a *   Contract (*   Edition dated   );

(h)   The Schedule of Basic Plant Charges (*   revision dated   );

(i)   Other documents as listed below:

2 We enclose for your approval the enclosed documents, which shall be deemed to form part of our tender, listed below:

3 We have obeyed the rules about confidentiality of tenders and will continue to do so as long as they apply.

4 Subject to and in accordance with paragraphs 3 above and the terms and conditions contained or referred to in the documents listed in paragraphs 1 and 2, we offer to execute the such works referred to in the said documents as the Employer may require in consideration of payment by the Employer of the rates shown below, plus reimbursement by the Employer of Value Added Tax in accordance with Condition 28 (VAT).

5 We shall receive for work executed or materials or labour supplied in the proper carrying out of any such Order:

(1) for work (or labour and/or materials supplied) measured and valued in accordance with Condition 27(1)(a) and (b)(Valuation) of the General Conditions of Contract, payment in accordance with Condition 27(6)(a) – for the purpose of which the Contractor's percentage adjustments shall be:

for orders for which the total of all the **net updated value** of measured work, in all sections of the Schedule of Rates, falls within the sums of £0–£5,000

+Add

...............................%

Deduct

for orders for which the total of all the **net updated value** of measured work, in all sections of the Schedule of Rates, falls within the sums of £5,001–£25,000

+Add

...............................%

Deduct

for orders for which the total of all the **net updated value** of measured work, in all sections of the Schedule of Rates, exceeds £25,001

+Add

...............................%

Deduct

(2) for work (or supply) valued in accordance with Condition 27(1)(c)(Valuation) of the General Conditions of Contract, payment of the value so ascertained;

(3) for work (or supply) valued in accordance with Condition 27(1)(d)(Valuation) of the General Conditions of Contract, payment of the value ascertained in accordance with Daywork Schedule;

6 We agree that the net rates and prices printed in or deduced from the Schedule as updated and adjusted by the adjustments referred to in paragraph 5 above shall be deemed to include all costs in respect of local conditions, labour, materials, plant, equipment, tools, insurances, overheads and profit, and all obligations, liabilities and services described in the Contract, but not any element of Value Added Tax.

7 (1) We may wish to employ subcontractors for the following services:

Service	Name of subcontractor (if known)

(2) We agree to seek the prior approval of the PM before employing any subcontractor on the Works.

8 We agree that differences or questions arising out of or relating to the Contract shall be resolved in accordance with Condition 38 (Adjudication) of the General Conditions.

SIGNED by

for and on behalf of

Tel:

Fax:

Telex:

Date:

*Employer to delete inapplicable item before issuing tender documents.

+Tenderer to complete

MODEL FORM 4

CONTRACT DAYWORK SCHEDULE

1 These terms and conditions apply to daywork executed by the Contractor and subcontractors.

2 The appropriate DEFINITION OF PRIME COST OF DAYWORK as stated in the tender applies, subject to any amendments therein.

3 Hourly Base Rates for labour are to be computed in accordance with Section 3.2 of the Definition and are to be inclusive of all incidental costs, overheads and profit as defined in Section 6 of the Definition. They are to be current at the month in which the due date for return of tenders falls, hereafter referred to as the Base Month, and will be subject to no further adjustment apart from updating as follows:

Hourly labour rate to be applied = $\quad HBR \times \dfrac{I_D}{I_O}$

Where HBR   =   Hourly Base Rate inserted at paragraph 7(1).

I_D =   The appropriate DETR INDEX for MTC DAYWORK (Classified Firm) labour index number for the calendar month during which daywork was executed as published by DETR in the Updating Percentage Adjustments for Measured Term Contracts.

I_O =   The appropriate labour index number for the Base Month.

4 Hourly Base Rates for operatives not listed at paragraph 7(1) shall be deduced by the Employer from the rates inserted against those listed.

5 The rates are to be calculated without any addition for Value Added Tax. Where applicable, VAT shall be calculated and reimbursed separately.

6 We agree that Section 1.2 of the Definition shall not apply and the rates in the Schedule of Basic Plant Charges as adjusted by percentage quoted at paragraph 7(3) shall apply to Plant used for works of a jobbing or maintenance character when ordered on Daywork under the Contract.

7   For work ordered to be carried out on Daywork, we require payment by the Employer as follows:

(1)   For Labour:

The Hourly Base Rate for the particular operative as stated below updated in accordance with paragraph 3 above. In the case of Plumbing Operatives, the National Working Rules of the Joint Boards for Plumbing Mechanical Engineering Services shall apply and Hourly Base Rates shall include for contributions to the relevant pension schemes.)

Hourly Base Rates+

£

Building Labour	(unskilled)
Building Labour	(skilled)
Plumbing Labour	(all grades)
Electrical Labour	(all grades)
Heating and Ventilating Labour	(all grades)

(2)   For materials and goods:

The cost as defined in Section 4 of the Definition, but after the deduction of any discount with the addition of   +per cent for incidental costs, overheads and profit as defined in Section 6 of the Definition.

(3)   For plant:

The cost as defined in Section 5 of the Definition calculated in accordance with the Schedule of Basic Plant Charges for Use in Connection with Daywork Under a   *   Contract, published by the RICS with the addition/deduction+ of   +per cent for incidental costs, overheads and profit as defined in Section 6 of the Definition.

(4)   Authorised overtime on daywork which has been worked on the written instruction of the PM shall be reimbursed as follows:

Productive Time:	As normal daywork detailed herein.
Non-Productive Time: (being the enhancements of time laid down in the Working Rule Agreement for the Appropriate trade)	As normal daywork detailed herein multiplied by 0.5

SIGNED by

for and on behalf of

Tel:

Fax:

Telex:

Date:

*Employer to delete inapplicable item before issuing tender documents.

+Tenderer to complete

MODEL FORM 5

ADJUDICATOR'S APPOINTMENT
(CONDITION 39)

THIS AGREEMENT is made the   day of

BETWEEN:

(1)

    of

    ('the Employer', which term shall include its successors and assignees);

(2)

    [of] OR [whose registered office is at]

    ('the Contractor'); and

(3)

    of

    ('the Adjudicator').

WHEREAS:

(A) The Employer has entered into a contract dated   **('the Contract')**
     with the Contractor for the execution of certain Works, and a copy of the Contract has been
     supplied to the Adjudicator.

(B) The Adjudicator has agreed to act as [adjudicator] OR [named substitute adjudicator] in
     accordance with the Contract.

NOW THIS AGREEMENT WITNESSETH as follows:

1   The Adjudicator shall, as and when required, act as [adjudicator] OR [named substitute adjudicator] in
    accordance with the Contract, except when unable so to act because of facts or circumstances beyond his
    reasonable control.

2   The Adjudicator confirms that he is independent of the Employer, the Contractor, and the Project Manager under the Contract, and undertakes to use reasonable endeavours to remain so. He shall promptly inform the Employer and the Contractor of any facts or circumstances which may cause him to cease to be so independent.

3   The Adjudicator hereby notifies the Employer and the Contractor that he will comply with Condition 39 (Adjudication) of the Contract, and its time limits.

4   The Adjudicator shall be entitled to take independent legal and other professional advice as reasonably necessary in connection with the performance of his duties as adjudicator. The reasonable net cost to the Adjudicator of such advice shall constitute expenses recoverable by the Adjudicator under this Agreement.

5   The Adjudicator shall comply, and shall take all reasonable steps to ensure that any persons advising or aiding him shall comply, with the Official Secrets Act 1989 and, where appropriate, with the provisions of Section 11 of the Atomic Energy Act 1946. Any information concerning the Contract obtained either by the Adjudicator or any person advising or aiding him is confidential, and shall not be used or disclosed by the Adjudicator or any such person except for the purposes of this Agreement.

6   The Employer and the Contractor shall pay the Adjudicator fees, expenses and other sums (if any) in accordance with the Contract and the Schedule, plus applicable Value Added Tax.

7   The Adjudicator is not liable for anything done or omitted in the discharge or purported discharge of his functions as adjudicator, unless the act or omission is in bad faith. Any employee or agent of the Adjudicator is similarly protected from liability.

8   The proper law of this Agreement shall be the same as that of the Contract. Where the proper law of this Agreement is Scots law, the parties prorogate the non-exclusive jurisdiction of the Scottish courts.

IN WITNESS whereof the Employer, the Contractor and the Adjudicator have executed this Agreement in triplicate on the date first stated above.

SCHEDULE

Adjudicator's Fees, Expenses, etc..

SIGNED by [name of signatory] for and on
behalf of [name of Employer]:

**SIGNED by [name of signatory] for and
on behalf of [name of Contractor]:**

SIGNED by [name of Adjudicator]:

MODEL FORM 6

ORDER TO PROCEED
(CONDITION 8)

Employer: *[insert name and address]*
Project: *[insert short description]*
Contract No:
Order No:
Contractor: *[insert name and address]*

To: the Contractor

Date:

Condition 1 (Definition, etc.): Description of the Works

The Contractor is hereby required to proceed with the following Works:

[insert description of the Works including drawings which numbered and attached with the Order form]

Condition 1 (Definition, etc.): Programme

The Contractor is required to commence work on *[insert date]* and complete by *[insert completion date]*

Condition 1 (Definition, etc.): Site

The Works are located at *[insert details of the Site]*

Condition 10 (Statutory Notices and CDM Regulations)

The CDM Regulations do/do not* apply to this Order.

Condition 20 (Passes)

Passes are/are not* required on connection with this Order.

Condition 29 (Advances on account)

The estimated value of the Works described in this Order is [insert estimated value exclusive of VAT] which is/is not* in excess of the figure stated in the Abstract of Particulars. The Contractor is/is not therefore entitled to claim advances on account.

and/or

The programme for the Works described in this Order exceeds/does not exceed 45 calendar days. The Contractor is/is not* entitled to claim advances on account.

This Order to Proceed is given under Condition 8 (Orders and Instructions).

....
Project Manager

*Delete inapplicable items

I acknowledge receipt of the above Order on

………………………………………..

Contractor's agent

MODEL FORM 7

INTERIM PAYMENT CERTIFICATE
(CONDITIONS 29 AND 31)

Employer: *[insert name and address]*
Project: *[insert short description]*
Project No:
Contract No:
Contractor: *[insert name and address]*

To: the Employer

Copied to: the Contractor

Date:

It is hereby certified under Condition 31 (Certifying payments) that the net sum not previously certified (taking into account all set-off or abatement to which the Employer is entitled, but exclusive of VAT) to which the Contractor is entitled under Condition 29(Advances on account) is   pounds (£   ), calculated on the following basis:

Condition 29 (Advances on account)

         £

Advance payment applied for by the
Contractor's detailed account (not
Previously certified)   …………………………...

Certified sum   ……………………………

…………………………...
Project Manager

MODEL FORM 8

FINAL PAYMENT CERTIFICATE
(CONDITION 30 and 31)

Employer: *[insert name and address]*
Project: *[insert short description]*
Project No:
Contract No:
Contractor: *[insert name and address]*

To: the Employer

Copied to: the Contractor

Date:

It is hereby certified under Condition 29 (Advances on account) that the net sum not previously certified (taking into account all set-off or abatement to which the Employer is entitled, but exclusive of VAT) to which the Contractor is entitled under Condition 30 (Final payment) is   pounds (£   ), calculated on the following basis:

Condition 30 (Final payment)   £

Final payment applied for by the
Contractor's detailed final account   ………………………………

Less amounts previously certified
under Condition 29 (Advances on account)   ………………………………

Sub-total   ………………………………

Certified sum   ………………………………

………………………………
Project Manager

MODEL FORM 9

NOTICE OF INTENTION TO WITHHOLD PAYMENT (CONDITION 32)

Employer: *[insert name and address]*

Project: *[insert short description]*

Project No:

Contract No:

Contractor: *[insert name and address]*

To: the Contractor

Date:

Notice is hereby given that the Employer proposes to withhold payment of   pounds
(£   ) upon the ground that *[insert details]*

OR

Notice is hereby given that the Employer proposes to withhold payment of a total of
pounds (£   ) upon the grounds that *[insert details of each ground for withholding and the amount proposed to be withheld attributable to each ground]*

This notice is given under Condition 32 (Payment notification and withholding payment).

....

Project Manager

MODEL FORM 10

EMPLOYER'S NOTICE OF DETERMINATION (CONDITION 36)

Employer: *[insert name and address]*
Project: *[insert short description]*
Project No:
Contract No:
Contractor: *[insert name and address]*

To: the Contractor

Date:

Notice is hereby given that the Contract is hereby determined, upon the ground mentioned in Condition 36(2)*(a)(b)(c)(d)(e)(f) (Determination by Employer), in that *[insert details]*

This notice is given under Condition 36(1) (Determination by Employer).

....
Project Manager

*Delete inapplicable items.

Barcellos Ltd

Sandbach House

8 Salisbury Road

Leicester LE1 7QR

Produce computerised MTC administration.

Building Cost Information Service and Building Maintenance

(Information service of the Royal Institution of Chartered Surveyors)

12 Great George Street

Parliament Square

London SW1P 3AD

http://www.bcis.co.uk

Carillion Services Ltd

Westlink House

981 Great West Road

Brentford

Middlesex TW8 9DN

Tel: (020) 8380 5323

Fax: (020) 8380 5070

http://carillionplc.com

Produce *PSA Schedules of Rates for Building Works,* eighth edition.

Housing Software Services

89 Gleneagle Road

London SW16 6AZ

Tel: (020) 8677 2253

Fax: (020) 8677 2101

http://www.hssl.co.uk

Produce *National Housing Federation Schedule of Rates.*

NSR Management Ltd

Rycote Place

30–38 Cambridge Street

Aylesbury HP20 1RS

Produce *National Schedule of Rates.*

QUDOS Computer Software Ltd

Ashmead House

3, The Common

Siddington

Cirencester GL7 6EY

Produce computerised MTC administrion.

Tudorseed Construction Ltd

Unit 3, Ripon House

35 Station Lane

Hornchurch

Essex RM12 6JL

Tel: (01708) 444678

Produce *Updating Percentages.*

Carillion Services Publications

Schedules of Rates

PSA Schedules of Rates can be used for both estimating building works and as a basis for contract. The rates are updated monthly by the 'Updating Percentage Adjustments for Measured Term Contracts'. The following publications are all published by The Stationery Office.

Costs-in-Use Tables

Essential for designers who need to take account of costs-in-use when selecting materials and components. The tables cover the more common forms of construction and provide data on all aspects of cleaning, maintenance and remedying defects.

ISBN 0-11-671531-6, 1991, £45

Schedule of Rates for Building Works

Essential as a price guide to tendering for contractors, who can use it as a basis for making quotations, and for clients, who can use it to judge the fairness of a quotation. Covers individual tasks involved in all building works.

ISBN 0-11-702400-7, 2000, £225

Schedule of Rates for Decoration Work

An authoritative source of reference, dealing solely with decorating. Includes a wide range of painting specifications and associated rates.

ISBN 0-11-702157-1, 1999, £45

Schedule of Rates for Electrical Services

Covers all aspects of electrical work, including specialist jobs such as the installation of fire alarm and public address systems, and the provision of aviation and ground lighting. Also includes rates for testing and inspection.

ISBN 0-11-702855-X, 2001, £185

Schedule of Rates for Landscape Management

Covers all aspect of landscape management and grounds maintenance. Ideal for use as a tendering and contract document for both lump sum and measured term contracts for new works and maintenace.

ISBN 0-11-702515-1, 2001, £120

Schedule of Rates for Mechanical Services

Offers comprehensive coverage for this specialist field, including installation of boilers, oil storage, air conditioning and thermal insulation, as well as pipework, valves, pumps and heaters.

ISBN 0-11-702856-8, 2001, £185